Cutting Corners

QUILTS WITH STITCH-AND-TRIM TRIANGLES

Joan **Hanson**

Martingale®
& COMPANY

Credits

President • Nancy J. Martin
CEO • Daniel J. Martin
Publisher • Jane Hamada
Editorial Director • Mary V. Green
Managing Editor • Tina Cook
Technical Editor • Ellen Pahl
Copy Editor • Karen Koll
Design Director • Stan Green
Illustrator • Laurel Strand
Text and Cover Designer • Trina Stahl
Photographer • Brent Kane

Acknowledgments

To Dawn Kelly of Sagle, Idaho: Thank you so much for your machine quilting skill and artistry. You make each quilt a one-of-a-kind beauty with a personality all its own. Not only that, you quilted 10 quilts in two weeks! Thank you!

To the staff at Martingale & Company: Thank you for editing instructions that are clear and easy-to-follow, designing books that are beautiful to look at, taking photographs that make every quilt look like a showstopper, and creating the marketing program that gets books to quilters who love them. I have enjoyed working with each of you, friends and quilt lovers all.

Cutting Corners: Quilts with Stitch-and-Trim Triangles
© 2003 by Joan Hanson

That Patchwork Place® is an imprint of Martingale & Company®.

Martingale & Company
20205 144th Avenue NE
Woodinville, WA 98072-8478
www.martingale-pub.com

Mission Statement

Dedicated to providing quality products and service to inspire creativity.

Printed in China
08 07 06 05 04 03   8 7 6 5 4 3 2 1

Library of Congress Cataloging-in-Publication Data

Hanson, Joan
   Cutting corners : quilts with stitch-and-trim triangles / Joan Hanson.
      p.   cm.
   ISBN 1-56477-474-0
1. Patchwork—Patterns.   2. Quilting.   3. Triangle.   I. Title.
TT835.H33625 2003
746.46'041—dc21

                                         2003004808

Contents

Introduction

Quilting should be fun. Too many people tell me, "I'd like to make quilts, but I don't have the time or patience." If they only knew the endless enjoyment they are missing out on!

My intent in writing this book is to present simple yet appealing quilts that are easy for beginning quilters to make, and for veteran quilters to make quickly. Good beginner projects are made with just squares and rectangles: Rail Fence, Log Cabin, Trip around the World, and Irish Chain, to name a few blocks. Then, sooner or later, blocks with triangles call out to be made: Sawtooth Star, Churn Dash, Snowball, Bow Tie, and Flying Geese.

Dividing a simple square into two triangles opens up a whole new world of design possibilities. The triangles are called half-square triangles. Clever quilters have devised many methods for making half-square triangles. The simplest method involves cutting a square ⅞" larger than the desired finished square, cutting it in half, and sewing it with another triangle the same size. Another method includes drawing grids to mark stitching and cutting lines. In yet another method, alternating colors of bias strips are sewn together and then cut into squares. All have their place and supporters. I have used them all and taught classes in many of them. I have found, however, that the cutting corners technique used in this book is one of the easiest for new quilters to grasp and use successfully.

This basic technique involves four very easy steps:

1. **Cut.** Cut squares as directed.

2. **Sew.** Place the square on the corner of a second piece of fabric and sew a diagonal line from corner to corner.

3. **Trim.** Trim off the excess fabric to a ¼" seam allowance.

4. **Press.** Fold over the triangle and press. Voilà!

Using this simple method, you can make as few or as many triangles as you want out of as few or as many different fabrics as you want. Sewing the diagonal seam before cutting results in a much more accurate half-square triangle, since the bias is more stable when contained within the square.

Taking this concept a step further, many more design possibilities emerge. What happens when four small squares are added to the corners of a large square? A Snowball block. Or two squares added to the ends of a rectangle? Flying Geese, or, looked at another way, star points. Now consider adding triangles to Four Patch or Nine Patch blocks and what do you get? An Hourglass block or Shoo Fly block. Making the Four Patch or Nine Patch block first and then adding the triangles to the corners gives you a much more accurate result. There are endless ways to use this very versatile and accurate technique.

In this book you will find a gallery of 35 blocks with instructions for constructing them using the cutting-corners technique. In addition, I have designed and included 15 easy yet very appealing projects that use loads of variations on the basic concept.

As you've most likely noticed, when triangles are made in this way, the corners are trimmed off, creating small triangles called waste triangles. Some are too small to worry about. Give them to kids to make collages, or put them out in your garden in the spring to add a bit of color to your mulch. Being a frugal person by nature, I am compelled to use the larger ones. The waste triangles that are trimmed from 2½" or 3½" squares can be used for other projects. Several projects with suitable waste triangles have companion projects that use up your scraps and waste triangles. I call these Waste Not, Want Not projects. Since they use up fabric you have sitting right in front of you and they are very quick to make, it is almost like making a second quilt for free.

The quilts are organized from the easiest to the more challenging, so pick one that matches your skill level, grab some fabrics, and start having some fun!

Quiltmaking **Basics**

Rotary Cutting

Good rotary-cutting equipment allows you to cut far more rapidly and with greater accuracy than with scissors. Purchase a cutter with a 2" or larger blade for quilting projects. Check the instructions that come with your cutter to see the proper way to hold your cutter and especially how to use the safety guard.

Safety First

Whether you are using rotary equipment for the first time or are already an old pro, get into the habit of following a few precautions. The rotary blade is extremely sharp, and you can unintentionally cut something important, like yourself. Make the following safety rules a habit whenever you use your rotary cutter.

▲ Always push the blade guard into place whenever you finish your cut. Keep the nut tight enough so that the guard won't slide back unintentionally.

▲ Keep your cutter in a safe place when not in use, especially if you have small children.

▲ Always roll the cutter away from you.

Joan's Helpful Hint

When your cutter starts to drag, oil the blade. Unscrew the nut on the back and lay out the pieces in order. Gently wipe the lint off the blade and guard with a scrap of fabric. Place a drop of sewing-machine oil on the side of the blade that touches the guard and reassemble the cutter. If the blade starts to skip, or becomes too dull to cut easily, replace it. Place the oiled side of the new blade toward the guard.

Cutting Strips

Cutting strips at an exact right angle to the folded edge of your fabric is the backbone to cutting accurate strips. This starts with the first cut, known as the clean-up cut.

1. After prewashing your fabric, fold it in half with the selvages together and press. Place the fabric on your cutting mat with the selvages toward you, and the fold even with a horizontal line at the top of the cutting mat. Place a 6" x 24" acrylic ruler so that the raw edges of both layers of fabric are covered, and the lines of your ruler match up with the vertical grid on your mat. Rolling the cutter away from you, cut from the selvages to the fold. Remove the ruler and gently remove the waste strip.

   To keep the ruler from shifting slightly or slipping as you cut, try cutting about 8" or 10" and then walking your hand forward on the ruler before cutting the next few inches. As soon as the cutter reaches the fabric next to your fingertips, stop and walk your hand forward on the ruler again.

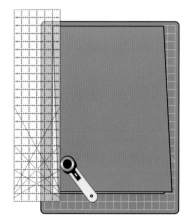

2. To make additional cuts, align the desired strip width measurement on the ruler with the cut edge of the fabric. Use the grid lines on your mat to double-check that you are making accurate cuts. After cutting three or four strips, realign the fold of your fabric with the lines on your mat and make a new clean-up cut.

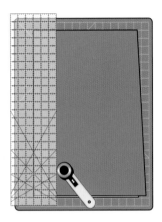

Sewing Basics

Set the stitch-length dial on your sewing machine to about 12 stitches per inch (written as 2.5 on many machines), and use a good-quality cotton thread.

Mastering the ¼" Seam

Sewing an exact ¼" seam is the key to successful quilt-making. A presser foot with the right-hand edge of the foot exactly ¼" from the needle makes it much easier. Your hands can keep the fabric feeding through the machine along the edge of the presser foot. Marking your sewing machine with ¼" masking tape as a sewing guide is also helpful. See "Diagonal Stitching Guide" on page 13 for instructions to mark your machine for the cutting corners technique.

Chain Sewing

This is a simple technique that will save time and thread.

1. Organize all the pieces that are to be joined with right sides together in a stack. Position the stack so that the side to be sewn is on the right. Start with the same edge on each pair and the same color on top; this will help you avoid confusion.

2. Sew each fabric pair one at a time without cutting the threads in between. As you add patches, your work will begin to look a bit like flags at an amusement park.

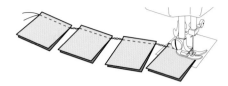

3. Take the chain of pieces to the ironing board, snip the pairs apart, and press them. You will find it tremendously satisfying to sew in chains and to watch as the parts of the blocks start to take shape on your ironing board.

Pressing

Precise piecing is a combination of accurate sewing and gentle pressing. Frequent light pressing enables you to see where the pieces should be matched. The important point to remember is to press frequently but lightly.

The traditional rule is to press seams to one side, toward the darker color whenever possible. Side-pressed seams add strength to the quilt, evenly distribute the bulk of the fabrics, and prevent the darker fabrics from showing through the lighter ones. Press the seam on the wrong side first; then turn the piece over and press from the right side.

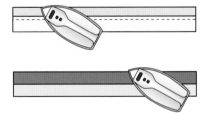

Pressing arrows are included when it is necessary to press the seams in a particular direction. When no arrows are indicated, the direction of the seam allowance doesn't matter.

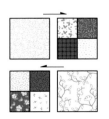
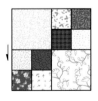

Assembling the Quilt Top

What a satisfying feeling to have finished making all your blocks! Now it is time to arrange them into rows and put the quilt top together. The best way to arrange the blocks is on a design wall. A design wall can be a piece of flannel or thin batting tacked to a wall, or it can be one or two sheets of foam-core board covered with flannel.

After arranging the blocks, sashings, setting squares, side and corner triangles, or other elements of your quilt, stand back and squint at your arrangement (or use a reducing glass). Are the colors evenly distributed? Does it match the quilt diagram? Are you pleased with how it looks? If the answers to these questions are all yes, good! If not, rearrange the pieces until you are pleased.

Joan's Helpful Hint

When you sew the rows together, remove one row from the design wall at a time, replacing the row when it is stitched together. It is easy to get confused as to what row you are working on.

Straight Settings

The blocks in a straight setting are arranged in straight rows and are easy to assemble.

1. Arrange the blocks in rows as shown in the quilt assembly diagram for your project, adding sashings and/or setting squares if applicable.

2. Sew the blocks together in horizontal rows. Press the seams in opposite directions from one row to the next, unless instructed otherwise. For example, if you have two different blocks that alternate with each other in your quilt, press all the seams toward the same block; or if you have sashings in your quilt, press all the seams either toward or

away from the sashings. Pressing arrows in the diagrams indicate the direction to press the seams.

3. Carefully pin the rows of blocks together, matching the seams. Stitch the rows together, and press the seams all in one direction.

Diagonal Settings

The blocks for diagonal settings are placed on point and arranged in diagonal rows. Corner and side setting triangles are then added to fill in the side and corner spaces.

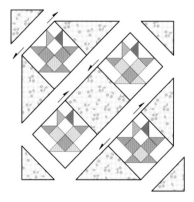

When blocks are set diagonally or on point, the bias grain runs horizontally and vertically, which can result in a quilt top that sags. To stabilize the quilt, cut the side and corner triangles so that the side of the triangles placed along the edge of the quilt are on the straight of grain.

In this book, the setting triangles are cut from slightly oversized squares. This makes sewing them to the quilt easier. The quilt top will then be trimmed before borders are added. The cutting directions for each quilt will tell you how large to cut these squares.

When you cut a square in half *once* on the diagonal, you create two triangles that have the straight of grain on the short sides. Sew these triangles to the corners of a diagonally set quilt.

When you cut a square in half *twice* on the diagonal, you produce four triangles with the straight of grain on the long side of each triangle. Sew these triangles along the side and top and bottom edges of a diagonally set quilt.

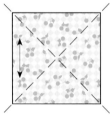 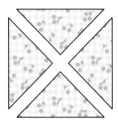

After assembling the diagonal rows of your quilt, you will need to trim the edges. Align the ¼" mark on your ruler with the points on the blocks and trim the edges ¼" from these points. In some cases, you will be instructed to trim more than ¼" away from the points ("Flying Geese Fishing Quilt"). This extra fabric between the points and the edge of the quilt is called "float"; it gives the appearance that the blocks are floating on the background.

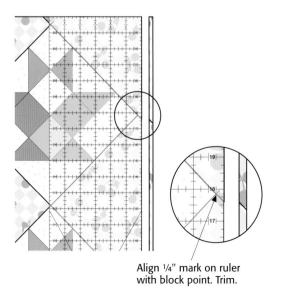

Align ¼" mark on ruler with block point. Trim.

Adding Borders

Adding borders can transform the look of a quilt. Quilt borders, like frames, can grab your eyes and focus them on the center blocks of a quilt. Borders emphasize and add significance to the designs they surround. Whether you add plain strips of fabric or rows of piecing, a thoughtfully planned border will greatly enhance a quilt. Well-designed borders echo the color, size, and shapes of the pieces in the blocks. The color of the outer border will bring out that color in the quilt. The best way to decide whether or not your quilt will benefit by a border is to experiment with some options. Use a design wall if you have one, or lay the quilt out flat on a bed or floor to audition border fabrics.

The single most important guide to remember about borders is that they must be cut to fit the actual size of the center of the quilt top, not the outer edges. It is important that the quilt end up "square" with 90° corners and with opposite sides equal to each other. In the cutting instructions for the projects, I have included an extra 3" for the border strips. Later, you will trim them to fit the exact measurements of the center of your quilt.

Joan's Helpful Hint

If borders are cut cross grain and must be pieced to obtain the length needed, cut off the selvage edges and join the strips with a straight seam.

Straight-Sewn Borders

Measure and cut border strips following the steps on page 9. By encouraging the quilt to fit the measured strips, you ensure that the quilt will be square with flat borders. This is an important step, so resist the temptation to skip it. You may have to ease one side of a quilt to fit a border and stretch the opposite side slightly to fit the same dimension. Press all border seams away from the center of the quilt top unless instructed otherwise.

1. Measure the length of the quilt through the center. Trim two of the border strips to that measurement. Sew these strips to the sides of the quilt, easing as necessary.

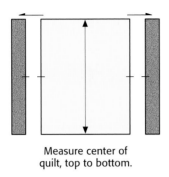

Measure center of quilt, top to bottom.

2. Measure the width of the quilt through the center, including the side borders and seam allowances as shown. Cut the borders to that length and sew them to the top and bottom of the quilt, easing or stretching as necessary.

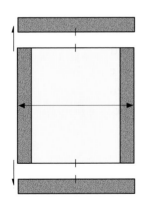

Measure center of quilt, side to side, including borders.

Straight-Sewn Borders with Corner Squares

1. Measure the length of the quilt through the center. Trim two of the border strips to that measurement. Measure the width of the quilt through the center and trim the top and bottom border strips to this measurement.

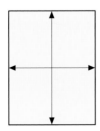

2. Sew the strips to the sides first, easing as necessary.

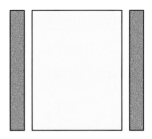

3. Sew a corner square to each end of the top and bottom strips, and sew these to the quilt, easing as necessary.

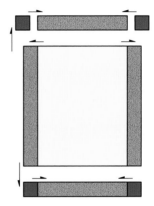

Finishing the Quilt

Decide which option you prefer to use for quilting your project: hand quilting, machine quilting on your home sewing machine, or machine quilting by a professional. If you want to have your quilt quilted by a professional, ask at your local quilt shop or quilt guild for a recommendation. Check with that person as to how you should prepare the top. Quilts usually do not need to be layered and basted for long-arm machine quilting, nor do they need to be marked.

Marking the Quilting Lines

Whether or not you mark quilting designs on the quilt top depends upon the type of quilting you will be doing. If you plan to outline quilt (quilt in the ditch), marking is not necessary. For more elaborate quilting designs, mark the quilt top before the quilt is layered with batting and backing.

Choose a marking tool that will be visible on your fabric and test it on fabric scraps to be sure the marks can be removed easily. If you are hand quilting, you can use ¼" masking tape for marking straight-line quilting after the quilt is basted.

The Backing

Cut a quilt backing that is at least 4" to 6" larger than your quilt top (2" to 3" all around). If your quilt is larger than the standard width of fabric, you will have to stitch two or more pieces of fabric together to make the backing. Trim away the selvages before piecing the lengths together, and press the seams open to make quilting easier. You can piece the backing either horizontally or vertically, whichever requires the least fabric.

Batting

Batting is available by the yard or packaged in standard bed sizes. Several weights are available. Thick battings are fine for tied quilts and comforters. However, a thinner batting is better if you intend to quilt by hand or machine.

Layering and Basting

Spread the backing wrong side up on a clean, flat surface. Use masking tape to anchor the backing to the surface without stretching the fabric. Spread the quilt batting on the backing, making sure it covers the entire backing and is smooth. Center the pressed and marked top on the batting and backing, right side up. Align borders and straight lines of the quilt top with the edges of the backing. Pin the layers together along the edge with large straight pins to hold the layers smooth.

For hand quilting, baste the three layers together, using a long needle and light-colored thread. Starting at the center, use large running stitches to baste across the quilt from side to side and top to bottom. Continue basting, creating a grid of parallel lines 6" to 8" apart. Complete the basting with a line of careful stitches around the outside edges.

For machine quilting, baste the layers with #2 rustproof safety pins. Place the pins about 6" to 8" apart, avoiding areas you intend to quilt.

Hand Quilting

Quilting is simply a short running stitch that goes through all three layers of the quilt. Hand-quilt in a frame, in a hoop, on a tabletop, or on your lap. Use a heavy thread designed for hand quilting. The thicker thread is less likely to tangle than regular sewing thread. Use a short, sturdy needle (called a Between) in a size #7. As you become more familiar with hand quilting, you will find that a smaller needle (higher number) will enable you to take smaller stitches. Use a thimble with a rim around the top edge to help push the needle through the layers.

1. Cut the thread 24" long and tie a small knot. Starting about 1" from where you want the quilting to begin, insert the needle through the top and batting only. Bring the needle up where the quilting will start. Gently tug on the knot until it pops through the quilt top and catches in the batting.

2. Insert the needle and push it straight down through all the layers. Then rock the needle up and down through all layers, "loading" three or four stitches on the needle. Push the needle with a thimble on your middle finger; then pull the needle through, aiming toward yourself as you work. Place your other hand under the quilt to make sure the needle has penetrated all three layers with each stitch. Continue in this way, taking small, even stitches through all three layers.

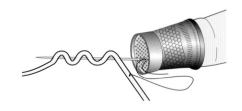

3. To end a line of quilting, make a small knot close to the quilt top and then take one stitch through the top and batting only. Pop the knot through the fabric into the batting. Clip the thread near the surface of the quilt.

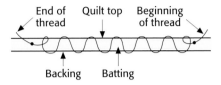

Machine Quilting

As quilters find that they are able to piece more quilt tops than they can quilt by hand, they realize that machine quilting is a practical alternative to hand quilting. In the last few years, machine quilting has become a beautiful art form in its own right. Choose a small quilt for your first machine-quilting project. A small project will be easier to guide through your sewing machine. Plan a quilting design that involves continuous, long, straight lines and gentle curves. Keep the spacing between quilting lines consistent over the entire quilt. Avoid tight, complex, little designs and don't leave large spaces unquilted.

Use a fine 100%-cotton, silk-finish thread or a very fine, high-quality .004 nylon thread made specifically for machine quilting. Nylon thread comes in clear for lighter fabrics and a smoke color for darker fabrics. Thread your bobbin with fine 100%-cotton thread.

Modern sewing machines have several features that are useful for machine quilting. Many machines have a walking foot or even-feed foot either built into the machine or as an attachment. This foot works on top of the fabric to move the top layer at the same speed as the bottom layer. This moves the quilt layers through the sewing machine evenly, avoiding puckering. Use this type of foot for straight-line quilting, grid quilting, and large, simple curves.

Walking Foot

Use a darning foot for curved designs and stipple quilting. This allows free fabric movement under the foot of your sewing machine. This is called free-motion quilting and, with practice, you can use it to produce beautiful quilting designs quickly. Choose designs that have continuous lines and that don't require a lot of starting and stopping. Lower the feed dog on your machine when quilting with a darning foot. This allows you to guide the fabric under the needle as if the needle were a stationary pencil. If you have a "needle down" feature on your machine, it is useful for all machine quilting. To start and stop, tie off your thread by shortening the stitch length for the first and last ⅛" to ¼".

Darning Foot

Trimming and Squaring Up

Once the quilting is complete, remove the basting thread or pins. Trim the batting and backing even with the quilt top, using a square ruler at the corners to make sure your quilt is as square as possible.

Making a Hanging Sleeve

1. If you are going to hang your quilt, attach a sleeve or rod pocket to the back before you bind the quilt. From the leftover backing fabric, cut a piece the width of your quilt by 8". On each end, fold under a ½" hem and then fold under again.

½" ½"

2. Fold the strip in half lengthwise, right sides together, and machine baste the raw edges to the top edge of your quilt. The sleeve will be stitched in place when you add the binding. Make a small pleat in the sleeve to accommodate the thickness of the rod and slipstitch the bottom edge of the sleeve edge to the backing fabric.

Sleeve

Quilt back

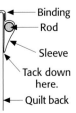

Binding
Rod
Sleeve
Tack down here.
Quilt back

Binding the Quilt

For a French double-fold binding, cut strips 2" to 2½" wide across the width of the fabric. A 2" binding strip will give you a binding that finishes at a scant ¼"; a 2½" binding strip will finish at about ⅜". You will need enough strips to go around the perimeter of the quilt, plus 10" to allow for turning the corners and for seams to join strips.

1. Sew the binding strips together to make one long strip. Join the strips at right angles, right sides together, and stitch across from corner to corner as shown. Trim the seam allowance to ¼" and press the seams open.

2. Fold the strip in half lengthwise, wrong sides together, and press. Turn under ¼" at a 45° angle at one end of the strip and press. Turning the end under at an angle distributes the bulk so you won't have a lump where the two ends of the binding meet.

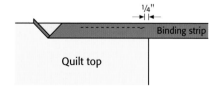

3. Starting on one side of the quilt and using a ¼"-wide seam allowance, stitch the binding to the quilt, keeping the raw edges even with the edge of the quilt top. Stop your stitching ¼" from the corner of the quilt and backstitch.

4. Remove the quilt from the sewing machine. Fold the binding back at a 45° angle away from the quilt and then down, so that it is even with the next side as shown. Start stitching at the edge, using a ¼" seam allowance as before.

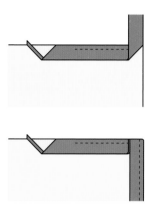

5. Continue stitching around all four sides, catching the sleeve on the top edge. When you get back to where you started, lap the end over the beginning and trim off the excess at an angle.

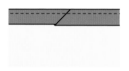

6. Fold the binding over the raw edges of the quilt to the back of the quilt, with the folded edge covering the row of machine stitching. Blindstitch the binding in place, mitering the corners.

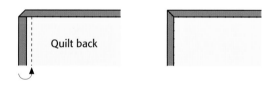

Adding a Quilt Label

Labels are an important finishing touch. A label can be as simple or as elaborate as you wish. Use a plain fabric that coordinates with your backing fabric. Include the name of the quilt, your name, your city and state, date, the recipient if the quilt is a gift, and any other interesting or important information. This can be embroidered or cross-stitched or written with a permanent pen. If you are using a pen, iron freezer paper to the back of the fabric to stabilize the fabric before writing.

Triangle Techniques

The stitch-and-trim or cutting corners method can be used for making half-square triangles, Snowball blocks, and flying-geese units accurately and easily. With this technique, you can also magically create Hourglass blocks from Four Patch blocks or a Shoo Fly block from a Nine Patch block simply by adding folded corners. I find it is more accurate to make many blocks with this method because you sew the straight lines first and then stitch the squares on the corners to create the corner triangles. You never have to cut triangles or sew along unstable bias edges. Use the four-patch-with-corners technique on page 18 to construct the Hourglass block. For the Shoo Fly block, make a Nine Patch block and use the snowball technique on page 15.

Read through the instructions beginning on page 14 for the basic half-square-triangle technique and then check out all the variations that follow. The directions for each quilt project in the book refer you back to these techniques as they are used in the quilts.

There are almost an unlimited number of blocks that can be made using this method. See the block gallery beginning on page 21 once you have familiarized yourself with the basic techniques.

Diagonal Stitching Guide

The first step in making easy and accurate diagonal seams used in the cutting-corners technique is to mark a stitching guideline on your sewing machine. You will make a stitching guide that you can use to line up with the machine needle. This line is used as a stitching guide in the same way that the edge of your presser foot or other guide is used to give you a ¼" seam allowance. In fact, if you use ¼" masking tape, you can mark both of these useful guidelines at the same time!

1. You will want to mark a guideline on your sewing surface extending at least 6" directly in front of your needle. (See "Extending Your Sewing Surface" on page 14 if you don't have 6" of surface area in front of your needle.) With the presser foot in the up position and the needle in the down position, line up a 6" x 12" ruler (or other large ruler) with the left side of the needle; the ruler should extend toward you in a straight line. Lower the pressure foot onto the ruler and temporarily tape the edges to secure the ruler in place.

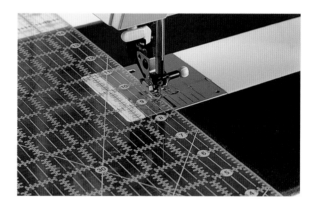

2. Starting in front of the presser foot, carefully run a strip of ¼" masking tape along the edge of the ruler to the front of the sewing surface.

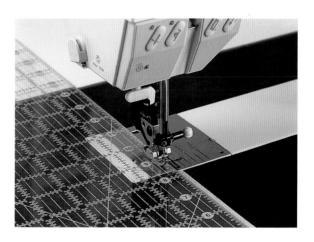

3. Remove the ruler and you have a center line and a ¼" seam allowance line marked on your sewing machine.

Think of this new guide in the same way as your ¼" seam guide, only this time you line up the diagonal points of your square with the needle and your center line guide (the left edge of the masking tape). Now you keep your eye on the bottom corner of the square traveling along the center guideline. With a little practice you will train your eye to steer your pieces, sewing an accurate diagonal line from corner to corner without marking.

Extending Your Sewing Surface

If you have a free-arm or portable sewing machine, you may need to improvise a sewing surface between you and your sewing machine. Look around your house and find a shoebox, videotape cases, a paperback book you've already read, or something that is about 5" wide and matches the height of your sewing-machine bed. Secure this to the front of your sewing machine with tape and follow steps 1 through 3 to make a stitching guideline with masking tape.

Half-Square Triangle: The Basic Cutting-Corners Technique

For all of the techniques, you begin by cutting a square that is ½" larger than the desired finished size of your triangle along the short edge. Each square when stitched will become one triangle after trimming and pressing.

1. **Cut.** Cut the squares as directed in the project instructions. Or determine the finished size of the outside edge of the triangle (short side); add ½" to the finished size to determine the size to cut the squares. Cut the squares to this size.

Short side

Finished size + ½" = cut size

2. **Sew.** Line up the squares right sides together, matching the raw edges exactly. Starting with the needle at the point of one corner and the opposite corner lined up with the center line guide, sew a diagonal line from corner to corner. Place your fingertips on each side of the fabric as you sew and keep your eyes on the bottom edge as it moves along the center line. If you have lots of half-square triangles to make, chain piece them one after the other.

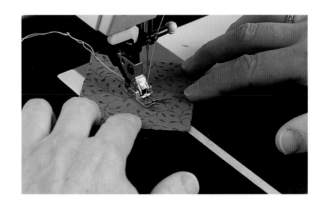

3. **Trim.** Using a small square ruler such as the Bias Square® (a 6" square rotary-cutting ruler) or the 4" Baby Bias Square® (I like this one because it is easy to hold on to), line up the ¼" line on your stitching line and trim off the corner of the squares so that you have a ¼" seam allowance remaining.

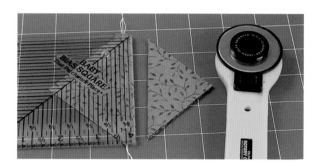

4. **Fold and press.** Gently press the seam allowance to one side, usually toward the darker fabric. To expedite pressing a large quantity, stack the pieces with the dark fabric on top. Peel back the dark fabric and guide the iron into the seam.

There are many variations of this basic technique that are used in the projects and blocks in this book. Here are the ones you'll need, along with some tips and tricks for each.

Snowball

1. Cut the large and small squares as directed in the pattern instructions. Place the small square on the corner of the large square, lining up the edges and outside corner.

2. Sew from corner to corner of the small square.

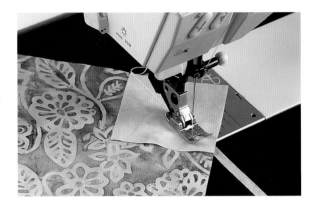

3. Trim off the corner, leaving a ¼" seam allowance.
4. Press toward the darker fabric.

Sew Quick

If you are making just a few Snowball blocks and want to sew more than one corner of the block before pressing, loosen but don't cut your thread. Place the next small square on the next corner of the big square and continue stitching. When all the corners are stitched, cut all the threads. Then trim and press all the corners at the same time.

If you are making multiples of the same block, chain piece one corner of all the blocks, cut the chain apart, sew the next small square to the next corner, and chain piece again. Repeat for each corner of the Snowball block. Then trim and press all the corners at the same time.

Flying Geese

This basic unit is used in many blocks and is super easy to make using this method.

1. Cut the rectangle and squares as directed in the pattern instructions. Place the square on the corner of the rectangle, lining up the edges.
2. Begin sewing on the long side of the rectangle. This helps keep the needle from pushing the fabric down the bobbin hole. Sew from corner to corner of the square. If you are making multiples, chain piece all the units.

3. Trim the corner off, leaving a ¼" seam allowance, and press.
4. Place the second square on the rectangle and position the unit to the right of the needle as shown so you can again stitch from the long side of the rectangle.

5. Trim and press.

Square within a Square

1. Cut the large and small squares as directed in the pattern instructions. Place the small square on the corner of the large square, lining up the edges.

2. Sew a diagonal line from corner to corner of the small square.

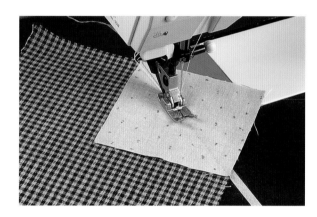

Sew Quick

When the finished triangles come to a point on the side of the center square, one triangle must be trimmed and pressed before the second triangle is sewn. Chain piece the first corner on all the blocks, and then the opposite corner of all the blocks. Cut the chain apart, trim, and press. Repeat for the opposite corners.

3. Place a second small square on the opposite corner of the large square, lining up the edges. Stitch from corner to corner.

4. Trim the corners off, leaving a ¼" seam allowance. Press.

5. Place the third small square on one of the remaining corners of the large square, lining up the edges. Stitch from corner to corner. Repeat for the fourth small square.

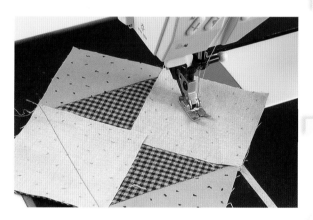

6. Trim and press.

Four Patch with Corners

Sewing a four-patch unit first and then adding the squares to the corners is much easier and more accurate than cutting and stitching eight triangles together to make this block.

1. Make the four-patch units as instructed in the project directions.

2. Follow steps 1 through 6 for the Square within a Square on page 17.

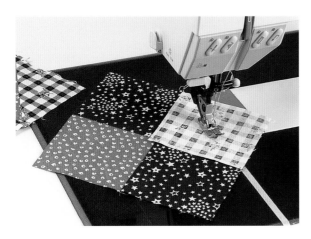

Rectangles with Diagonal Seams

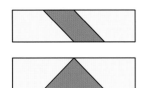

Here's another great way to use the cutting corners technique. You will start by cutting three rectangles.

1. Cross the first two rectangles, lining up the corners as shown. Since you can't see the corner of the rectangle underneath from the top, mark the corner on the top rectangle with a pencil or a pin.

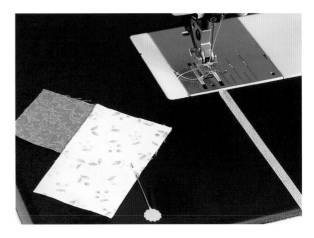

2. Begin sewing at the middle of the rectangle underneath in the same way as the flying-geese units. Stitch the diagonal seam, using the pin or pencil mark as a guide.

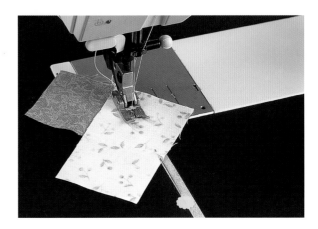

Sew Quick

Once you get the hang of this, you may not need to mark the hidden corner at all; you may be able to peek under the top piece and line up the bottom corner with your center line tape as you sew.

3. Trim the corner off, leaving a ¼" seam allowance, and press.

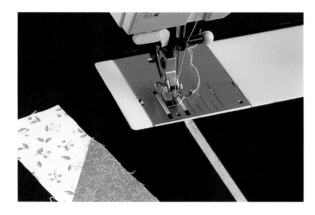

4. To make a rectangular unit with a parallelogram in the middle, align the third rectangle with the unit from step 3 and sew the next seam parallel to the first seam.

5. To make a rectangular unit with a triangle in the middle, sew the second seam in the opposite direction .

Big and Little Triangles

In this technique, you will create a large half-square triangle that adjoins the long sides of two smaller triangles, usually with a square in the opposite corner. In order to conserve fabric, start with a "phantom" corner unit that consists of three squares.

1. Place the large square on top of the pieced unit specified in the block or project instructions, lining up the two sides. Flip the unit over and locate the corner of the square on the pieced unit where the stitching line will cross. Mark this point with a pin.

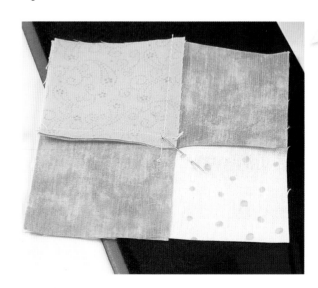

2. Flip the unit over so that the large square is on top. Sew the diagonal line, aiming for the pin, and then the corner point.

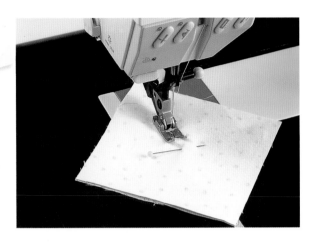

3. Trim the corner off, leaving a ¼" seam allowance. Press.

Using Waste Triangles

The leftover waste triangles from the 2½" squares and the 3½" squares are big enough to trim down to a smaller size and use in another project. I designed several projects that use these sizes. They follow the larger project and take advantage of other leftover fabric as well. When you plan to use the waste triangles in another project, trim them down to size right away while they are still lined up with each other. What a great feeling to have all the triangles cut for another project before you are done with the first one! Use a 6" Bias Square, 4" Baby Bias Square, or other square rotary-cutting ruler to trim the waste triangle to the desired size. I especially like the Baby Bias Square for trimming waste triangles; it is a perfect size for that purpose. There will be a missing or "phantom" corner, but this is the corner that will be trimmed again anyway. There will be enough of the two sides remaining for lining up the piece.

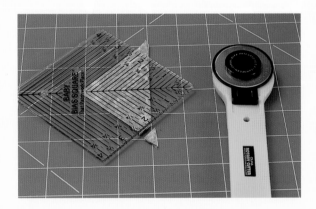

Trim the waste triangle to the desired "square" size.

Trim the seam allowance to ¼" after sewing on the diagonal.

Block **Gallery**

As you start looking at blocks and the way they are constructed, you will notice that many can be pieced using one of the cutting-corners techniques. To help get you started, I've included a gallery of some common blocks along with the cutting sizes and piecing instructions for the cutting-corners method.

Snowball

Piece	Number to Cut	Size
A	1	8½" x 8½"
B	4	2½" x 2½"

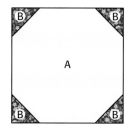

Finished Size: 8"

Empty Spool

Piece	Number to Cut	Size
A	1	3½" x 9½"
B	2	3½" x 9½"
C	4	3½" x 3½"

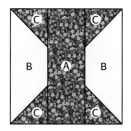

Finished Size: 9"

Interlocking Rings

Piece	Number to Cut	Size
A	1	4½" x 4½"
B	4	2½" x 2½"
C	4 light + 4 medium	1½" x 2½"
D	4	1½" x 3½"
E	8 medium + 4 dark	1½" x 1½"

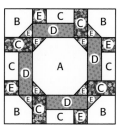

Finished Size: 8"

Tea Cup

Piece	Number to Cut	Size
A	1	4" x 4½"
B	1	2" x 6½"
C	2	1½" x 1½"
D	2	2" x 2"
E	2	1½" x 4"
F	1	1½" x 6½"
G	1	½" x 4" bias

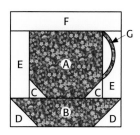
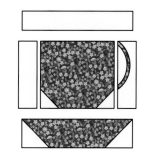

Finished Size: 6"

Bow Tie

Piece	Number to Cut	Size
A	2	3½" x 3½"
B	4 medium 1, 6 medium 2	2" x 2"

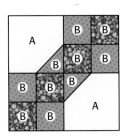
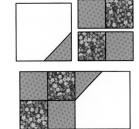

Finished Size: 6"

Mosaic

Piece	Number to Cut	Size
A	2 medium 1, 2 medium 2	3½" x 3½"
B	4 medium 3, 4 medium 4	2½" x 2½"
C	4 light, 4 dark	1½" x 1½"

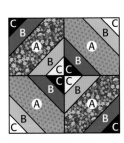

Finished Size: 6"

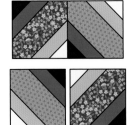

Churn Dash

Piece	Number to Cut	Size
A	4 light, 4 dark	3½" x 3½"
B	5 light, 4 medium	2" x 2"

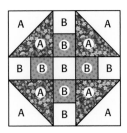
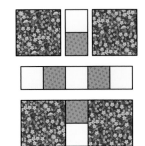

Finished Size: 7½"

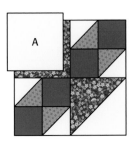

Old Maid's Puzzle

Piece	Number to Cut	Size
A	2 light, 2 dark	3½" x 3½"
B	4 light, 4 medium 1, 4 medium 2	2" x 2"

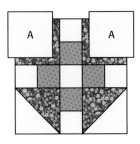
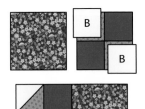

Finished Size: 6"

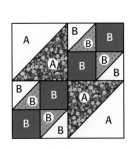

Road to Oklahoma

Piece	Number to Cut	Size
A	10 light, 6 medium, 4 dark	2" x 2"

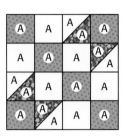 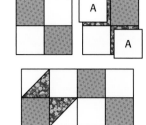

Finished Size: 6"

Sawtooth Star

Piece	Number to Cut	Size
A	1	4½" x 4½"
B	4	2½" x 4½"
C	4 light, 8 dark	2½" x 2½"

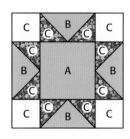 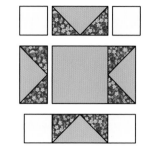

Finished Size: 8"

Flying Geese

Piece	Number to Cut	Size
A	3	3½" x 6½"
B	6	3½" x 3½"
C	2	2" x 9½"

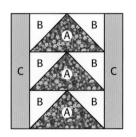 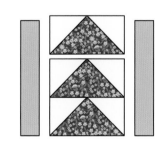

Finished Size: 9"

Heart

Piece	Number to Cut	Size
A	2	3½" x 6½"
B	2	3½" x 3½"
C	4	2" x 2"

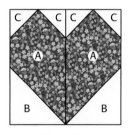 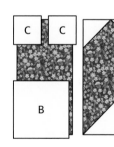

Finished Size: 6"

King's Crown

Piece	Number to Cut	Size
A	1	4½" x 4½"
B	4	2½" x 4½"
C	4 light, 8 medium	2½" x 2½"

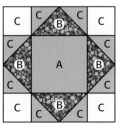

Finished Size: 8"

Prairie Flower

Piece	Number to Cut	Size
A	1 medium, 4 light, 4 dark	3½" x 3½"
B	8 light, 8 medium	2" x 2"

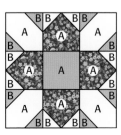 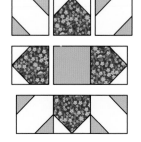

Finished Size: 9"

Big T

Piece	Number to Cut	Size
A	4 light, 5 medium	3½" x 3½"
B	4 light, 4 dark	2" x 3½"
C	8	2" x 2"

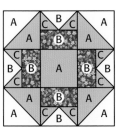
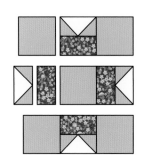

Finished Size: 9"

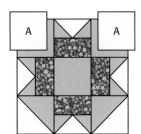

Berkeley

Piece	Number to Cut	Size
A	1 medium, 4 light	3½" x 3½"
B	8	2" x 3½"
C	12	2" x 2"

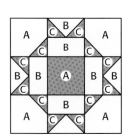

Finished Size: 9"

Seesaw

Piece	Number to Cut	Size
A	4 light, 4 dark	2½" x 4½"
B	4 medium, 8 light	2½" x 2½"

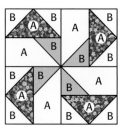
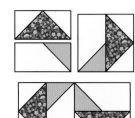

Finished Size: 8"

Louisiana

Piece	Number to Cut	Size
A	4 light, 4 dark	2½" x 4½"
B	8	2½" x 2½"

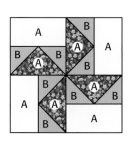
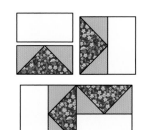

Finished Size: 8"

Duck Paddle

Piece	Number to Cut	Size
A	4 medium, 4 dark	3½" x 3½"
B	8	2" x 3½"
C	4 light, 17 medium	2" x 2"
D	4	2" x 5"

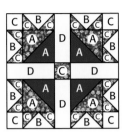

Finished Size: 10½"

Double Star

Piece	Number to Cut	Size
A	1 light, 4 medium	3½" x 3½"
B	12	2" x 3½"
C	16 light, 20 medium	2" x 2"

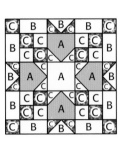

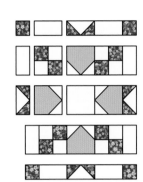

Finished Size: 12"

Robbing Peter to Pay Paul

Piece	Number to Cut	Size
A	1	3½" x 3½"
B	4 light, 4 medium	2" x 3½"
C	12 light, 4 medium, 20 dark	2" x 2"

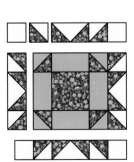

Finished Size: 9"

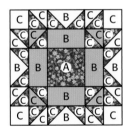

Memory

Piece	Number to Cut	Size
A	1 medium	4½" x 4½"
B	4 light, 4 dark	2½" x 4½"
C	12 light, 16 medium, 12 dark	2½" x 2½"

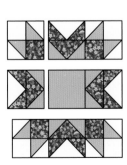

Finished Size: 12"

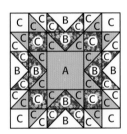

Square within a Square

Piece	Number to Cut	Size
A	1 medium 1, 4 medium 2	4½" x 4½"
B	4	2½" x 4½"

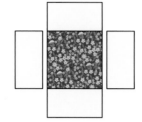
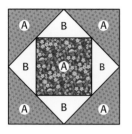

Finished Size: 8"

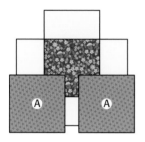
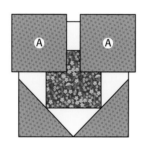

Double Square within a Square

Piece	Number to Cut	Size
A	1 light, 4 medium	4½" x 4½"
B	4 dark, 4 light	2½" x 2½"
C	4	2½" x 4½"

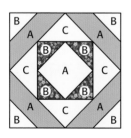
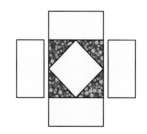

Finished Size: 8"

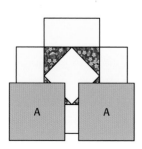
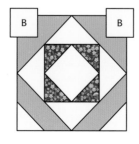

Variable Star

Piece	Number to Cut	Size
A	1	4½" x 4½"
B	4	2½" x 4½"
C	4 light, 4 medium, 8 dark	2½" x 2½"

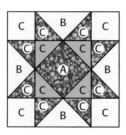
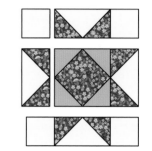

Finished Size: 8"

Union Square

Piece	Number to Cut	Size
A	5	3½" x 3½"
B	4 light, 4 medium	2" x 3½"
C	16 light, 12 medium	2" x 2"

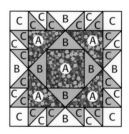
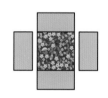

Finished Size: 9"

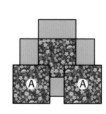
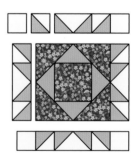

Rolling Stone

Piece	Number to Cut	Size
A	5	3½" x 3½"
B	4 light, 4 dark	2" x 3½"
C	16	2" x 2"

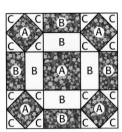

Finished Size: 9"

Wheel of Time

Piece	Number to Cut	Size
A	4 light, 8 medium, 12 dark	2½" x 2½"
B	4	2½" x 4½"

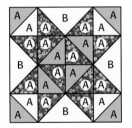

Finished Size: 8"

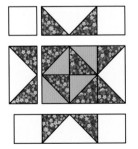

Connecticut

Piece	Number to Cut	Size
A	16 light, 6 medium, 2 dark	2½" x 2½"
B	4	2½" x 4½"

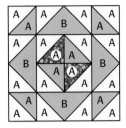

Finished Size: 8"

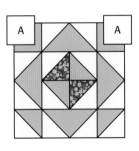
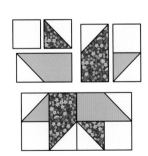

Judy's Star

Piece	Number to Cut	Size
A	2 medium, 2 dark	2½" x 4½"
B	12 light, 2 medium, 2 dark	2½" x 2½"

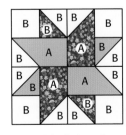
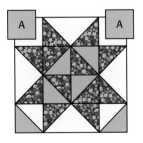

Finished Size: 8"

Checked Square

Piece	Number to Cut	Size
A	24 light, 6 medium, 6 dark	2" x 2"
B	4 medium, 4 dark	2" x 3½"

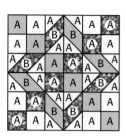

Finished Size: 9"

Carpenter's Star

Piece	Number to Cut	Size
A	1	3½" x 3½"
B	28 light, 4 medium	2" x 2"
C	12	2" x 3½"

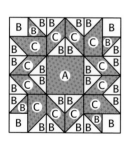

Finished Size: 9"

State Fair

Piece	Number to Cut	Size
A	1	3½" x 3½"
B	8	2" x 3½"
C	16 light, 4 medium, 8 dark	2" x 2"

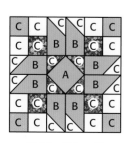

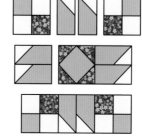

Finished Size: 9"

Basket

Piece	Number to Cut	Size
A	2 light, 5 medium, 4 dark	2½" x 2½"
B	4	2½" x 4½"
C	1	4½" x 4½"

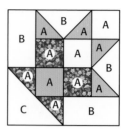
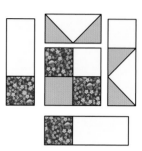

Finished Size: 8"

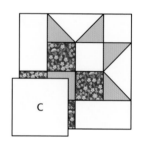

Blooming Star

Piece	Number to Cut	Size
A	4 medium 1, 4 medium 2, 1 dark	3½" x 3½"
B	16 light, 4 medium	2" x 2"

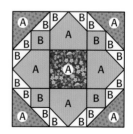

Finished Size: 9"

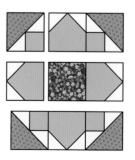

Batik **Nine Patch**

Designed and pieced by Joan Hanson, 58" x 75". Quilted by Dawn Kelly.

Walk into any quilt shop and you'll be confronted with an abundance of fabulous fabrics just waiting to inspire you to make them into beautiful quilts. Some luscious batik fabrics beckoned to me one day, and I thought how wonderful it would look to combine them with some hand-dyed fabrics I had. I chose a Snowball block to showcase the batik fabric and a Nine Patch block for the solid fabrics. I set the blocks on point, which makes them look more complex than they really are.

Finished Quilt Size: 58" x 75" ▲ Finished Block Size: 6"
Number of Blocks: 35 Nine Patch blocks and 24 Snowball blocks

Materials

42"-wide fabric

▲ 2½ yards of batik for Snowball blocks and outer border

▲ 2 yards of light background for blocks and setting triangles

▲ ¼ yard *each* of 7 medium hand-dyed fabrics for Nine Patch blocks*

▲ ¾ yard of dark hand-dyed fabric for Nine Patch blocks and inner border

▲ 3¾ yards fabric for backing

▲ ⅝ yard fabric for binding

▲ 64" x 81" piece of batting

With seven fabrics, you will make five blocks of each color. If you want a scrappier look, use more fabrics.

Cutting

Fabric	Used for	Number to cut	Size to cut	Second cut
Light background	Nine Patch blocks	14 strips	2½" x 42"	14 strips, 2½" x 27"
	Snowball blocks	2 strips	2½" x 42"	96 squares, 2½" x 2½", (use the leftover strips from the Nine Patch blocks also)
	Side setting triangles	5 squares	10" x 10"	Cut twice diagonally.
	Corner setting triangles	2 squares	5½" x 5½"	Cut once diagonally.
7 assorted medium hand-dyed fabrics	Nine Patch blocks	2 strips of each fabric	2½" x 42"	1 strip, 2½" x 27" 2 strips, 2½" x 14"
Dark hand-dyed fabric	Nine Patch block centers	3 strips	2½" x 42"	7 strips, 2½" x 14"
	Inner border	6 strips	2" x 42"	—
Batik	Snowball blocks	2 lengthwise strips	6½" x 80"	24 squares, 6½" x 6½"
	Outer border	2 lengthwise strips	6½" x 66"	—
		2 lengthwise strips	6½" x 61"	—

Piecing the Nine Patch Blocks

1. Sew a 2½" x 27" background strip to either side of a 2½" x 27" medium hand-dyed strip to make strip set A. Press the seams toward the dark fabric. Repeat with all seven of the solid fabrics. Cut each of the seven strip sets into 10 segments, each 2½" wide.

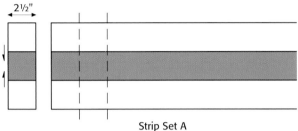

2½"

Strip Set A
Make 7. Cut 70 segments (10 from each strip set).

2. Sew a 2½" x 14" medium hand-dyed strip to either side of a 2½" x 14" dark hand-dyed strip to make strip set B. Press the seams toward the dark fabric. Repeat with all seven of the medium hand-dyed fabrics. Cut the strip sets into 5 segments, each 2½" wide, for a total of 35.

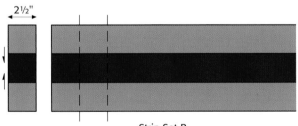

2½"

Strip Set B
Make 7. Cut 35 segments (5 from each strip set).

3. Sew a strip-set A segment to either side of a strip-set B segment, matching the solids in each Nine Patch block. Repeat to make five of each solid color. Press the seam allowances toward the center segments.

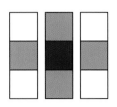 

Make 35.

Piecing the Snowball Blocks

Using the snowball technique on page 15, sew a 2½" x 2½" background square to each corner of a 6½" x 6½" batik square. Press toward the batik fabric. Repeat with all 24 batik squares and 96 background squares.

Make 24.

Assembling the Quilt

1. Lay out the Nine Patch blocks in diagonal rows, alternating them with the Snowball blocks and referring to the quilt diagram on page 32. Arrange the side and corner setting triangles as shown.

2. Sew the blocks and side setting triangles together in rows, pressing all seam allowances toward the Nine Patch blocks.

3. Sew the rows together, adding the corner triangles last. Press all seam allowances in one direction.

4. Square up the quilt, trimming the edges ¼" from the corners of the blocks.

Adding the Borders

1. Sew the dark hand-dyed inner-border strips together to make a long continuous strip.

2. Measure the length of your quilt, referring to "Adding Borders" on page 8, and cut two border strips to this length. Sew the border strips to the sides of the quilt and press the seam allowances toward the border strips.

3. Measure the width of your quilt top including the side inner-border strips, and cut two inner-border strips to this length. Sew the border strips to the top and bottom of your quilt and press the seam allowances toward the border strips.

4. Repeat steps 2 and 3 for the outer border. Since you cut the border strips lengthwise, piecing the border is unnecessary.

Finishing the Quilt

Refer to the general directions on pages 9–12 for more details on quilt finishing, if needed.

1. Piece the backing horizontally and cut the backing fabric so that it is approximately 4" to 6" larger than the quilt top.

2. Layer the backing, batting, and quilt top, and baste the layers together.

3. Hand or machine quilt as desired.

4. Trim the batting and backing fabric so that the edges are even with the quilt top. Add a sleeve for hanging the quilt, if desired. Bind and label your quilt.

Jewel **Box**

Designed and pieced by Joan Hanson, 59 ½" x 74 ½". Quilted by Dawn Kelly.

This block is sometimes called Buckeye Beauty and sometimes Jewel Box. Whichever name you prefer, it is very easy to make, with two four-patch units and two triangle units in each block. It is an ideal block for using up your favorite scraps, especially if you save strips and squares from other projects. I have used the three primary colors, red, blue, and yellow, as well as green. The blocks are mostly blues (21 blocks), with some reds and pinks (13 blocks), some greens (9 blocks), and just a few yellows (5 blocks). Choose a variety of lights and medium to darks in these four colors for your blocks. They can be as scrappy as your fabric stash allows. The fabric used for the inner border and sashing squares includes the four colors used in the blocks and helps unify the quilt. Note that the outside row of sashing squares features the blue outer-border fabric to contrast with the inner-border fabric.

Finished Quilt Size: 59½" x 74½" ▲ Finished Block Size: 6" ▲ Number of Blocks: 48

Materials

42"-wide fabric

- ▲ 2½ yards of dark blue for sashing squares and outer border
- ▲ 2 yards total of assorted light scraps for blocks
- ▲ 2 yards total of assorted medium-to-dark scraps for blocks
- ▲ 1¼ yards of light print for sashing

- ▲ ¾ yard of floral print for sashing squares and inner border
- ▲ 3¾ yards of fabric for backing
- ▲ ⅝ yard of fabric for binding
- ▲ 66" x 80" piece of batting

Cutting

Fabric	Used for	Number to cut	Size to cut	Second cut
Assorted light scraps	Blocks	10 strips	2" x 42"	—
	Blocks	9 strips	3½" x 42"	96 squares, 3½" x 3½"
Assorted medium-to-dark scraps	Blocks	10 strips	2" x 42"	—
	Blocks	9 strips	3½" x 42"	96 squares, 3½" x 3½"
Dark blue	Outer border	4 lengthwise strips	5¼" x 68"	—
	Sashing squares	2 strips	2" x 42"	28 squares, 2" x 2"
Light print	Sashing	19 strips	2" x 42"	110 rectangles, 2" x 6½"
Floral print	Inner border	6 strips	2" x 42"	—
	Sashing squares	2 strips	2" x 42"	35 squares, 2" x 2"

Piecing the Blocks

1. Sew a light 2" strip to a medium-to-dark 2" strip, keeping the color family the same (e.g., light blue and dark blue). Make 10 strip sets. Press the seam allowances toward the dark strip. Cut the strip sets into a total of 192 segments, each 2" wide.

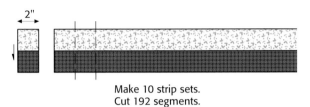

Make 10 strip sets.
Cut 192 segments.

2. Sew two segments of the same color family together to make a four-patch unit. Make a total of 96 units.

Make 96.

3. Sew two light squares and two four-patch units of the same color family together into a large four-patch unit as shown. Press the seam allowances toward the small four-patch units. Make 48 blocks.

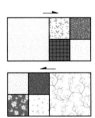 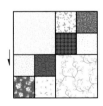

Make 48.

4. Using the four-patch-with-corners technique on page 18, sew two medium-to-dark squares of the same color to the corners as shown. Press the seam allowances toward the dark triangles. Repeat for all 48 blocks.

Make 48.

Piecing the Sashing

1. Sew seven dark blue squares and six light print rectangles together as shown. Repeat to make two sashing units. Press the seam allowances toward the squares.

Make 2.

2. Repeat step 1 using the dark blue squares on the outside of the unit and the floral-print squares for the inner squares as shown. Make seven sashing units.

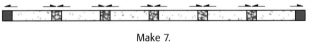

Make 7.

Assembling the Quilt

1. Lay out the blocks in rows, alternating six blocks and seven sashing strips as shown. Stitch together and press the seam allowances toward the blocks to avoid show-through of the seam allowances. Make eight rows, alternating the angle of the blocks from row to row.

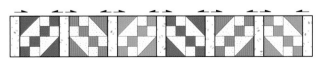

Make 8.

2. Sew the rows together, alternating with the sashing units and placing the two units with all dark blue squares at the top and bottom of the quilt. Press the seam allowances toward the block rows.

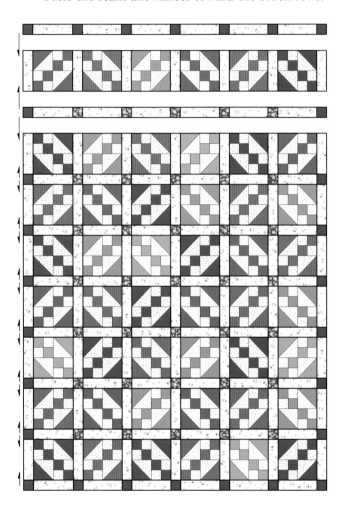

Adding the Borders

1. Sew the floral strips together to make one long inner-border strip.

2. Measure the length of your quilt, referring to "Adding Borders" on page 8, and cut two border strips to this length. Sew the borders to the sides of the quilt and press the seam allowances toward the border strips.

3. Measure the width of your quilt top, including the side inner-border strips. Cut two inner-border strips to this length. Sew the border strips to the top and bottom of your quilt and press the seam allowances toward the border strips.

4. Repeat steps 2 and 3 for the outer border.

Finishing the Quilt

Refer to the general directions on pages 9–12 for more details on quilt finishing, if needed.

1. Piece the backing fabric horizontally and cut it so that it is approximately 4" to 6" larger than the quilt top.

2. Layer the backing, batting, and quilt top, and baste the layers together.

3. Hand or machine quilt as desired.

4. Trim the batting and backing fabric so that the edges are even with the quilt top. Add a sleeve for hanging the quilt, if desired. Bind and label your quilt.

Garden **Path**

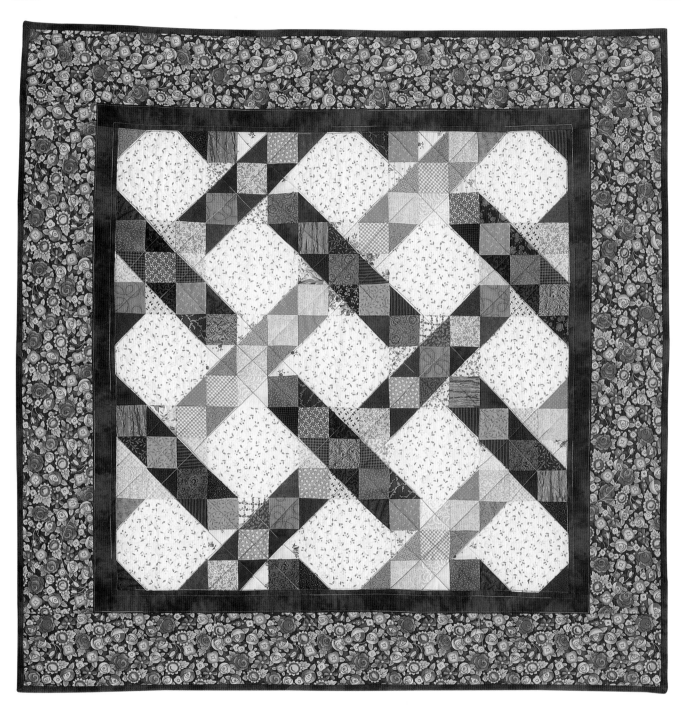

Designed and made by Joan Hanson, 41" x 41".

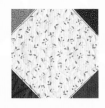

This fascinating quilt is actually made of two very simple blocks, the Snowball block and the Contrary Wife block. When set together, they create a wonderful illusion of colors weaving together. What a satisfying quilt to make out of leftover pieces.

Finished Quilt Size: 41" x 41" ▲ Finished Block Size: 6"
Number of Blocks: 13 Snowball blocks and 12 Contrary Wife blocks

WASTE NOT, WANT NOT

Use waste triangles from the quilt "Jewel Box" on page 33 or cut squares as directed below.

For Snowball Blocks
13 waste triangles *each* of dark red, blue, yellow, and green (52 total)
OR
Cut 52 squares, 2½" x 2½".

For Contrary Wife Blocks
12 pairs of dark and light waste triangles *each* of red, blue, yellow, and green (96 total)
OR
Cut 96 squares, 2½" x 2½".

NOTE: *If you run short of waste triangles of a particular color, fill in with 2½" x 2½" squares.*

Materials
42"-wide fabric

▲ ¾ yard of light print for Snowball blocks
▲ ⅝ yard of dark print for outer border
▲ ⅜ yard of dark blue for inner border
▲ ⅛ yard *each* or scraps of medium red, blue, green, and yellow for Contrary Wife blocks
▲ 2 yards of fabric for backing
▲ ½ yard of fabric for binding
▲ 47" x 47" piece of batting

Cutting

Fabric	Used for	Number to cut	Size to cut	Second cut
Light print	Snowball blocks	3 strips	6½" x 42"	13 squares, 6½" x 6½"
Medium scraps	Contrary Wife blocks	15 squares each of red, blue, yellow, and green	2½" x 2½"	—
Dark blue	Inner border	4 strips	2" x 35"	—
Dark print	Outer border	4 strips	4¼" x 42"	—

Piecing the Snowball Blocks

1. Referring to "Using Waste Triangles" on page 20, trim 13 waste triangles of each color (red, blue, yellow, and green) to 2½" x 2½" on two sides as shown.

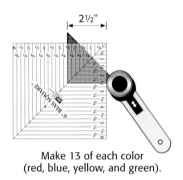

Make 13 of each color
(red, blue, yellow, and green).

2. Using the snowball technique on page 15, sew a dark color 2½" x 2½" waste triangle to each corner of the light print squares. Trim the seam allowances to ¼" and press toward the corners. Make nine with color combination A and four with color combination B.

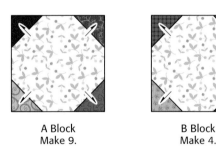

A Block
Make 9.

B Block
Make 4.

Piecing the Contrary Wife Blocks

1. Referring to "Using Waste Triangles" on page 20, trim 12 waste-triangle pairs of each color (12 dark and 12 light) down to 2½" x 2½" on two sides as shown. Using the half-square-triangle technique on page 14, sew the light and dark waste triangles together. Trim the seam allowances to ¼" and

press toward the dark triangles. Make 12 half-square triangles each of red, blue, yellow, and green for a total of 48.

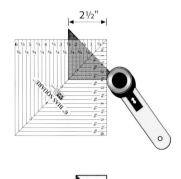

Make 12 of each color
(red, blue, yellow, and green).

2. Arrange the 2½" medium squares and the half-square triangles in the following color combinations. Sew the squares together in rows, pressing the seam allowances toward the plain squares. Sew the rows together. Make three blocks of each.

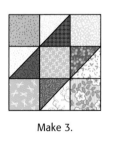

Make 3.

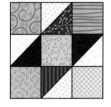

Make 3.

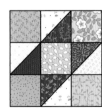

Make 3.

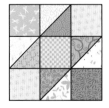

Make 3.

Assembling the Quilt

1. Referring to the quilt diagram, lay out the blocks in rows, alternating the Snowball blocks and Contrary Wife blocks so that the colors appear to weave together. Sew together in rows. Press the seam allowances toward the Contrary Wife blocks.

2. Sew the rows together. Press the seam allowances all in one direction.

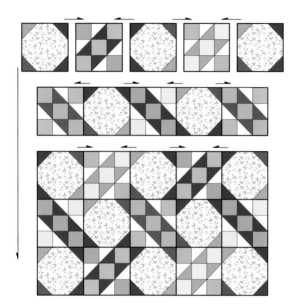

Adding the Borders

1. Measure the length of your quilt, referring to "Adding Borders" on page 8, and trim two of the dark blue border strips to this length. Sew the border strips to the sides of the quilt and press the seam allowances toward the border strips.

2. Measure the width of your quilt top, including the side inner-border strips, and trim the remaining two dark blue border strips to this length. Sew the border strips to the top and bottom of your quilt and press the seam allowances toward the border strips.

3. Repeat steps 2 and 3 for the outer border.

Finishing the Quilt

Refer to the general directions on pages 9–12 for more details on quilt finishing, if needed.

1. Cut the backing fabric so that it is approximately 4" to 6" larger than the quilt top.

2. Layer the backing, batting, and quilt top, and baste the layers together.

3. Hand or machine quilt as desired.

4. Trim the batting and backing fabric so that the edges are even with the quilt top. Add a sleeve for hanging the quilt, if desired. Bind and label your quilt.

Holly Berry **Wreath**

Designed and pieced by Joan Hanson, 53" x 53". Quilted by Dawn Kelly.

This Christmas wreath quilt is a perfect size to hang over a fireplace or sofa during the holiday season. It would also make a great table topper for a round or square table. Four red and four green print and plaid metallic fabrics from a fat-quarter packet were used for the center medallion and pieced border of Bow Tie blocks.

Finished Quilt Size: 53" x 53" ▲ Finished Block Size: 6" ▲ Number of Blocks: 28

Materials
42"-wide fabric

- ▲ 1½ yards of dark red for borders A, C, and E
- ▲ 1¼ yards of light print for background
- ▲ ½ yard or 2 fat quarters *each* of green 1 and 4 for center medallion and blocks
- ▲ 4 fat quarters of assorted red fabrics (red 1, 2, 3, and 4) for center medallion and blocks
- ▲ ⅝ yard of red-and-green print for border B
- ▲ 1 fat quarter *each* of green 2 and 3 for center medallion and blocks
- ▲ 3½ yards of fabric for backing
- ▲ ½ yard of fabric for binding
- ▲ 59" x 59" piece of batting

Joan's Helpful Hint

When pressing metallic fabrics, use a dry iron that is set to a medium setting. A hot setting may melt the fabric.

Piecing the Center Medallion

1. Using the snowball technique on page 15, sew two 3½" green 1 squares to opposite corners of the 12½" light print square. Sew the two 3½" x 3½" green 2 squares to the remaining two corners. Press seam allowances toward the green fabrics.

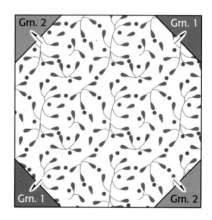

2. Sew a 3½" green 1 square to one corner of two 6½" light print squares. Repeat with two 3½" green 2 squares and the remaining two 6½" light print squares. Press toward the green fabric.

Make 2.

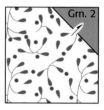
Make 2.

Cutting

Fabric	Used for	Number to cut	Size to cut	Second cut
Light print	Center medallion	1 square	12½" x 12½"	—
		4 squares	6½" x 6½"	—
		8 squares	3½" x 3½"	—
		2 strips	2" x 10"	—
	Bow Tie border blocks	5 strips	3½" x 42"	56 squares, 3½" x 3½"
Green 1	Center medallion	4 squares	3½" x 3½"	—
		3 strips	2" x 10"	—
	Bow Tie border blocks	6 strips	2" x 20"	56 squares, 2" x 2"
		3 strips	2" x 20"	—
Green 2	Center medallion	4 squares	3½" x 3½"	—
	Bow Tie border blocks	3 strips	2" x 20"	—
Green 3	Center medallion	4 squares	3½" x 3½"	—
		1 strip	2" x 10"	—
Green 4	Center medallion	4 squares	3½" x 3½"	—
		3 strips	2" x 10"	—
	Bow Tie border blocks	6 strips	2" x 20"	—
Red 1	Center medallion	2 strips	2" x 10"	—
	Bow Tie border blocks	3 strips	2" x 20"	—
Red 2	Center medallion	2 strips	2" x 10"	—
	Bow Tie border blocks	3 strips	2" x 20"	—
Red 3	Center medallion	2 strips	2" x 10"	—
	Bow Tie border blocks	3 strips	2" x 20"	—
Red 4	Center medallion	1 strip	2" x 10"	—
	Bow Tie border blocks	3 strips	2" x 20"	—
Dark red	Border A	4 strips	1½" x 42"	2 strips, 1½" x 24½"
				2 strips, 1½" x 26½"
	Border C	4 strips	1½" x 42"	2 strips, 1½" x 34½"
				2 strips, 1½" x 36½"
	Border E	6 strips	2¾" x 42"	—
Red-and-green print	Border B	4 strips	4½" x 42"	2 strips, 4½" x 26½"
				2 strips, 4½" x 34½"

3. Sew the 2" x 10" strips together in the following combinations. Press all the seam allowances toward the green fabrics or away from the light print fabric. Cut each strip set into four 2"-wide segments.

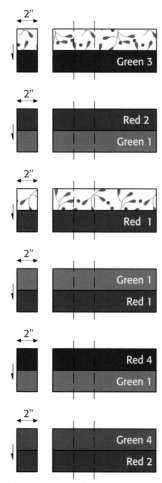

Make 1 strip set of each combination.
Cut 4 segments from each strip set.

Make 2 strip sets.
Cut 8 segments.

4. Sew the segments into four-patch units in the combinations shown. Make four of each.

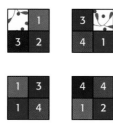

Make 4 of each.

5. Sew the four-patch units and 3½" green 3, green 4, and light print squares together as shown. Press. Make four units.

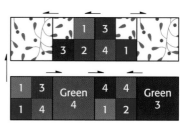

Make 4.

6. Sew the blocks and units together in three rows as shown. Sew the rows together to make the center medallion.

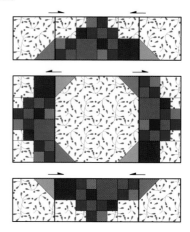

Piecing the Bow Tie Blocks

1. Using the snowball technique on page 15, sew a 2" green 1 square to a corner of a 3½" light print square. Press the seam allowances toward the green fabric. Make 56.

Make 56.

2. Sew the 2" x 20" strips together in the following combinations. Make three strip sets of each combination. Press all the seam allowances toward the green fabrics. Cut each strip set into 28 segments, each 2" wide.

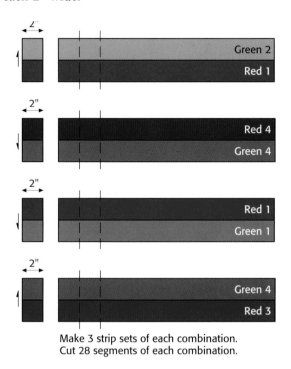

Make 3 strip sets of each combination.
Cut 28 segments of each combination.

3. Sew the segments into four-patch units in the combinations as shown. Make 28 of each.

Make 28.   Make 28.

4. Sew the Bow Tie blocks together as shown. Press the seam allowances toward the four-patch units. Make 28 blocks.

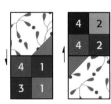 

Make 28.

Adding the Borders

1. Sew the two dark red strips, 1½" x 24½", to opposite sides of the center medallion. Press the seam allowances toward the red border. Sew the two 1½" x 26½" dark red strips to the remaining two sides of the center medallion.

2. Repeat step 1 to add the red-and-green print border strips. Then add the dark red border C strips. Your piece should now measure 36½" square so that the Bow Tie border will fit properly.

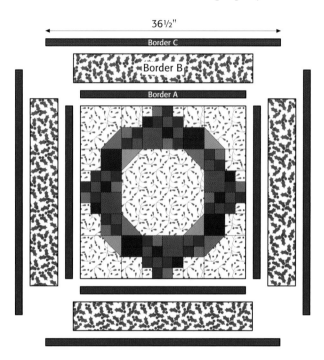

3. Sew six Bow Tie blocks together as shown. Make two units. Sew eight Bow Tie blocks together as shown; make two units.

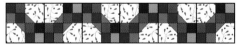

Make 2.

Make 2.

4. Sew the six-block units to opposite sides of the quilt. Press the seam allowances toward the red border. Add the eight-block units to the top and bottom of your quilt. Press.

5. Sew the 2¾" dark red border E strips together to make one long strip.

6. Measure the length of your quilt, referring to "Adding Borders" on page 8, and cut two border strips to this length. Sew the red border strips to the sides of the quilt and press the seam allowances toward the red border.

7. Measure the width of your quilt top, including the side borders, and cut two red border strips to this length. Sew the border strips to the top and bottom of your quilt and press the seam allowances toward the red border.

Finishing the Quilt

Refer to the general directions on pages 9–12 for more details on quilt finishing, if needed.

1. Piece the backing fabric vertically and cut so that it is approximately 4" to 6" larger than the quilt top.

2. Layer the backing, batting, and quilt top, and baste the layers together.

3. Hand or machine quilt as desired.

4. Trim the batting and backing fabric so that the edges are even with the quilt top. Add a sleeve for hanging the quilt, if desired. Bind and label your quilt.

Flying Geese **Fishing Quilt**

Designed and pieced by Joan Hanson, 63" x 76". Quilted by Dawn Kelly.

I made this quilt out of the soft flannels that are flooding into our fabric stores. The Flying Geese blocks are very easy and quick with large pieces that are perfect for using those wonderful flannel fabrics. The side and corner triangles show off a fishing theme print that provided the color palette for the other fabrics. This is the ideal quilt to curl up with in front of a fire on a cold winter night. Choose all flannels for your quilt, or choose regular quilting cottons.

Finished Quilt Size: 63" x 76" ▲ Finished Block Size: 9" ▲ Number of Blocks: 32

Materials

42"-wide fabric

- ▲ 1¾ yards of theme print for side and corner setting triangles
- ▲ 1¾ yards *total* of assorted dark scraps for A and B block large triangles
- ▲ 1¼ yards of cream print for background of A blocks
- ▲ 1⅛ yards of red print for border
- ▲ ⅞ yard of light print for background of B blocks
- ▲ ¾ yard of blue print for sides of A blocks
- ▲ ½ yard of green print for sides of B blocks
- ▲ 4½ yards of fabric for backing
- ▲ ⅝ yard of fabric for binding
- ▲ 68" x 81" piece of batting

Piecing the Blocks

1. Using the flying-geese technique on page 16, sew a cream square to each side of a dark rectangle. Press toward the dark rectangle. Make 60 for the A blocks. Repeat with the light print squares and the remaining dark rectangles. Make 36 for the B blocks.

Make 60.   Make 36.

2. Sew the triangle units into groups of three. Make 20 for the A blocks and 12 for the B blocks. Press the seam allowances toward the dark fabrics.

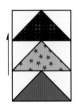   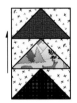

Block A   Block B
Make 20.   Make 12.

Cutting

Fabric	Used for	Number to cut	Size to cut	Second cut
Cream print	A blocks	11 strips	3½" x 42"	120 squares, 3½" x 3½"
Assorted dark scraps	A and B blocks	16 strips	3½" x 42"	96 rectangles, 3½" x 6½"
Light print	B blocks	7 strips	3½" x 42"	72 squares, 3½" x 3½"
Blue print	A blocks	10 strips	2" x 42"	40 rectangles, 2" x 9½"
Green print	B blocks	6 strips	2" x 42"	24 rectangles, 2" x 9½"
Theme print	Side triangles	4 squares	18" x 18"	Cut twice diagonally. *
	Corner triangles	2 squares	11" x 11"	Cut once diagonally.
Red print	Border	8 strips	4¼" x 42"	—

* You will need just 14 of these triangles.

3. Add a blue rectangle to each side of the A block units and a green rectangle to each side of the B block units. Press the seam allowances toward the strips.

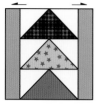
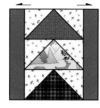

Block A   Block B

Assembling the Quilt

1. Lay out the A blocks and B blocks in diagonal rows, referring to the quilt assembly diagram below. Arrange the side setting triangles and corner triangles as shown.

2. Sew the blocks and side setting triangles together in rows, pressing all seam allowances toward the A blocks.

3. Sew the rows together, adding the corner triangles last. Press all seam allowances in one direction.

Assembly Diagram

4. Square up the quilt, trimming the edges 2" from the corners of the blocks.

Adding the Borders

1. Sew the red strips together in pairs to make four long border strips.

2. Measure the length of your quilt, referring to "Adding Borders" on page 8, and trim two of the border strips to this length. Sew the border strips to the sides of the quilt and press the seam allowances toward the border strips.

3. Measure the width of your quilt top, including the side borders, and trim the remaining two border strips to this length. Sew the border strips to the top and bottom of your quilt and press the seam allowances toward the border strips.

Finishing the Quilt

Refer to the general directions on pages 9–12 for more details on quilt finishing, if needed.

1. Piece the backing fabric vertically and trim it approximately 4" to 6" larger than the quilt top.

2. Layer the backing, batting, and quilt top, and baste the layers together.

3. Hand or machine quilt as desired.

4. Trim the batting and backing fabric so that the edges are even with the quilt top. Add a sleeve for hanging the quilt, if desired. Bind and label your quilt.

Joan's Helpful Hint

If you are using a directional print for the side and corner triangles, check to make sure they line up properly and are not upside down or sideways.

Pinwheel **Star**

Designed and made by Joan Hanson, 40" x 50".

There are almost 200 pairs of waste triangles from "Flying Geese Fishing Quilt," and these Pinwheel Star blocks use up about three quarters of them. Add a few scraps, some sashing strips, and a border, and this quilt practically makes itself!

WASTE NOT, WANT NOT

Use waste triangles from "Flying Geese Fishing Quilt" on page 47 or cut squares as directed below.

For Pinwheel Star Blocks

144 pairs of cream and dark waste triangles

OR

Cut 288 squares, 2½" x 2½"

(144 light and 144 dark).

NOTE: *You will need 12 pairs of the same color combination for each Pinwheel Star block.*

Materials

42"-wide fabric

▲ ⅞ yard of red print for border

▲ ¾ yard of light green for sashing strips

▲ ⅜ yard of assorted light cream prints for background squares

▲ Assorted scraps of dark fabric for sashing squares

▲ 2⅝ yards of fabric for backing

▲ ½ yard of fabric for binding

▲ 46" x 56" piece of batting

Piecing the Blocks

1. Referring to "Using Waste Triangles" on page 20, trim the pairs of waste triangles down to 2½" x 2½" on two sides as shown. Make 12 pairs of the same color combination for each Pinwheel Star block.

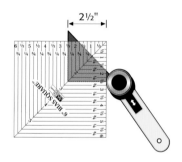

2. Using the half-square-triangle technique on page 14, sew the pairs together. Trim the seam allowances and press toward the dark fabric. Make 12 for each block.

Make 12
per block.

Cutting

Fabric	Used for	Number to cut	Size to cut	Second cut
Cream prints	Pinwheel Star block background	3 strips	2½" x 42"	48 squares, 2½" x 2½"
Dark scraps	Sashing squares	20 squares	2½" x 2½"	—
Light green	Sashing strips	8 strips	2½" x 42"	31 rectangles, 2½" x 8½"
Red print	Border	5 strips	4¼" x 42"	—

3. Arrange the units from step 2 and four cream squares in four rows as shown. Sew the units together in rows, pressing the seam allowances in opposite directions in each row.

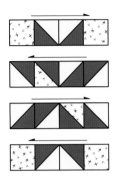

4. Sew the rows together. Press the seam allowances open to reduce bulk. Repeat to make 12 blocks.

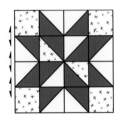

Make 12.

Assembling the Quilt

1. Arrange the blocks in rows, alternating with sashing strips and sashing squares as shown.

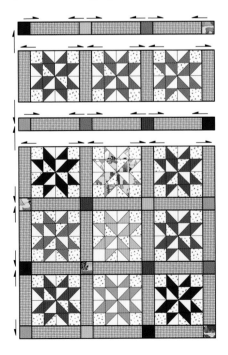

2. Sew the blocks and sashing strips into rows. Sew the rows together. Press all the seam allowances toward the sashing rows.

Adding the Borders

1. Measure the length of your quilt, referring to "Adding Borders" on page 8. Piece three red print border strips together, if necessary, and cut two strips to this length. Sew the border strips to the sides of the quilt and press the seam allowances toward the border strips.

2. Measure the width of your quilt top, including the side border strips, and trim the remaining two red border strips to this length. Sew the border strips to the top and bottom of your quilt and press the seam allowances toward the border strips.

Finishing the Quilt

Refer to the general directions on pages 9–12 for more details on quilt finishing, if needed.

1. Piece the backing fabric horizontally and trim so that it is approximately 4" to 6" larger than the quilt top.

2. Layer the backing, batting, and quilt top, and baste the layers together.

3. Hand or machine quilt as desired.

4. Trim the batting and backing fabric so that the edges are even with the quilt top. Add a sleeve for hanging the quilt, if desired. Bind and label your quilt.

Little Fishing **Quilt**

Designed and made by Joan Hanson, 28" x 28".

This little wall hanging would be just the thing for someone in your life who loves fishing. You can easily change the mood to fit the person by choosing a different theme fabric. It's perfect for hanging in an office, den, or family room. The quilt is quick to make and uses up a small fraction of the waste triangles from "Flying Geese Fishing Quilt." It's ideal for using fat quarters, too. If you want to make a larger version, increase the number of blocks to 36 and make the quilt 6 blocks across and 6 blocks down.

Finished Quilt Size: 28" x 28" ▲ Finished Block Size: 5" ▲ Number of Blocks: 16

WASTE NOT, WANT NOT

Use waste triangles from "Flying Geese Fishing Quilt" on page 47, or cut squares as directed below.

For Blocks
32 pairs of cream and dark waste triangles
OR
Cut 64 squares, 2½" x 2½".

Materials
42"-wide fabric

- ▲ ¾ yard of theme print for border
- ▲ ⅜ yard of red print for blocks and corner squares
- ▲ ¼ yard or scraps of green for blocks
- ▲ ⅛ yard or scraps of cream print for blocks
- ▲ ⅛ yard or scraps of gold print for blocks
- ▲ Scraps of dark prints and plaids for blocks
- ▲ 1 yard of fabric for backing
- ▲ ⅜ yard of fabric for binding
- ▲ 34" x 34" piece of batting

Cutting

Fabric	Used for	Number to cut	Size to cut	Second cut
Cream print	Blocks	16 squares	2½" x 2½"	—
Dark prints and plaids	Blocks	16 squares	2½" x 2½"	—
Green fabric	Blocks	2 strips	1½" x 42"	32 rectangles, 1½" x 2½"
Red print	Blocks	2 strips	1½" x 42"	32 rectangles, 1½" x 2½
	Corner squares	4 squares	4¼" x 4¼"	—
Gold print	Blocks	1 strip	1½" x 42"	16 squares, 1½" x 1½"
Theme print*	Border	4 strips	4¼" x 23"	—

Note: If your theme print is a directional fabric, cut two border strips crosswise and two lengthwise.

Piecing the Blocks

1. Using a Bias Square ruler, trim all the waste triangle pairs down to 2½" x 2½" on two sides, as shown.

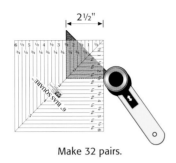

Make 32 pairs.

2. Using the half-square-triangle technique on page 14, sew the waste triangle pairs together. Trim the seam allowances and press toward the dark fabric. Make 32 half-square-triangle units.

Make 32.

3. Arrange the half-square-triangle units, cream squares, dark squares, red and green rectangles, and the small gold squares as shown to make the blocks. Sew the pieces together in three rows. Press the seam allowances toward the red and green rectangles. Make 16 blocks.

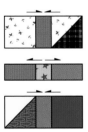 

Make 16.

Assembling the Quilt

1. Referring to the quilt diagram, lay out the blocks in rows to create the diamond pattern. Sew the blocks together in rows. Press the seam allowances in opposite direction from row to row.

2. Sew the rows together. Press the seam allowances all in one direction.

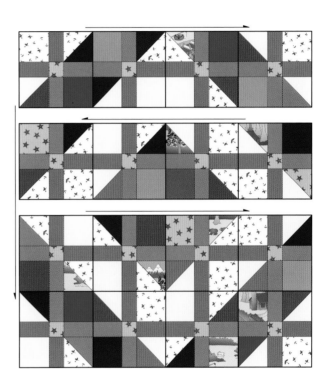

Joan's Helpful Hint

Before sewing the blocks together, try turning them in different directions. Many design possibilities will emerge right before your eyes.

Adding the Borders

1. Measure the length of your quilt, referring to "Adding Borders" on page 8, and trim all four of the border strips to this length. Sew two of the border strips to the sides of the quilt and press the seam allowances toward the border strips.

2. Sew the corner squares to the ends of the remaining two border strips. Press the seam allowances toward the borders.

3. Sew the border strips to the top and bottom of your quilt and press the seam allowances toward the borders.

Finishing the Quilt

Refer to the general directions on pages 9–12 for more details on quilt finishing, if needed.

1. Cut the backing fabric so that it is approximately 4" to 6" larger than the quilt top.

2. Layer the backing, batting, and quilt top; baste the layers together.

3. Hand or machine quilt as desired.

4. Trim the batting and backing fabric so that the edges are even with the quilt top. Add a sleeve for hanging the quilt, if desired. Bind and label your quilt.

Cherry Star **Picnic**

Designed and pieced by Joan Hanson, 58½" x 58½". Quilted by Dawn Kelly.

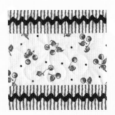

Look closely and you will see that this charming quilt is made up of three very simple blocks: a Star block, an Hourglass block, and a strip-pieced block. The strip-pieced block is the easiest, and you will make more of those than the other two. How easy can it get? Striped fabric and rickrack add a lot of energy and movement to this quilt without adding to the difficulty. Look for the giant rickrack in larger fabric stores where ribbons and trims are sold by the yard; it is larger than the packaged jumbo size.

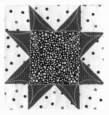

Finished Quilt Size: 58½" x 58½" ▲ Finished Block Size: 6"
Number of Blocks: 16 Star blocks, 9 Hourglass blocks, and 24 strip-pieced blocks

Materials

42"-wide fabric

▲ 2 yards of black-and-white check for Hourglass blocks and outer border
▲ 1⅝ yards of light cherry print for strip-pieced blocks and inner border
▲ ¾ yard of light fabric or scraps for background of Star blocks
▲ ⅝ yard *total* of assorted red scraps for Star and Hourglass blocks
▲ ⅝ yard *total* of assorted black scraps for Star and Hourglass blocks

▲ ⅝ yard of red-and-white stripe for strip-pieced blocks
▲ ⅜ yard of red fabric for middle border
▲ 3⅝ yards of fabric for backing
▲ ⅝ yard of fabric for binding
▲ 64" x 64" piece of batting
▲ 10 yards of black narrow rickrack for strip-pieced blocks
▲ 6 yards of black giant rickrack for border

Cutting

Fabric	Used for	Number to cut	Size to cut	Second cut
Light	Star block background	9 strips	2" x 42"	64 rectangles, 2" x 3½" 64 squares, 2" x 2"
Assorted red and black scraps	Star block centers	16 squares	3½" x 3½"	—
	Star block points	128 squares	2" x 2"	—
Assorted black scraps	Hourglass blocks	18 squares	3½" x 3½"	—
Assorted red scraps	Hourglass blocks	18 squares	3½" x 3½"	—
Black-and-white check	Hourglass blocks	36 squares	3½" x 3½"	—
	Outer border	4 lengthwise strips	5¾" x 61"	—
Red-and-white stripe	Strip-pieced blocks	8 strips	2" x 42"	—
Light cherry print	Strip-pieced blocks	4 lengthwise strips	3½" x 42"	—
	Inner border	4 lengthwise strips	2½" x 51"	—
Red	Middle border	5 strips	1¼" x 42"	—

Piecing the Star Blocks

1. Sew a 2" red or black square to each end of a 2" x 3½" background rectangle using the flying-geese technique on page 16. Press the seam allowances toward the red and black star points. Make 4 units of the same color combination for each Star block for a total of 64 units.

Make 64.

2. Sew a star-point unit to opposite sides of a 3½" red or black square. Press the seam allowances toward the star points. Make 16 units.

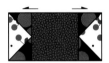

Make 16.

3. Sew a background square to opposite sides of the remaining star-point units. Press the seam allowances toward the star points. Make 32 units.

Make 32.

4. Sew the units from steps 2 and 3 together as shown to make the blocks. Press the seam allowances toward the center of the block. Make 16 blocks.

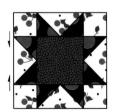

Make 16.

Piecing the Hourglass Blocks

1. Sew two red squares and two black squares together to make a four-patch unit. Press the seam allowances toward the black squares. Make nine four-patch units.

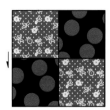

Make 9.

2. Sew a black-and-white-check square to opposite corners using the four-patch-with-corners technique on page 18. Press the seam allowances toward the check fabric. Add black-and-white-check squares to the remaining two corners. Press the seam allowances toward the black-and-white-check. Make nine blocks.

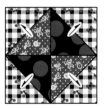

Make 9.

Making the Strip-Pieced Blocks

1. Using a 1" seam-allowance guide on your sewing machine, sew the narrow rickrack to the center of the red-and-white strips, using a thread color that matches the rickrack.

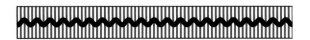

2. Sew a red-and-white strip to either side of the 3½"-wide cherry-print strips to make four strip sets. Press the seam allowances toward the red-and-white strips. Cut the strip sets into a total of 24 blocks, each 6½" wide.

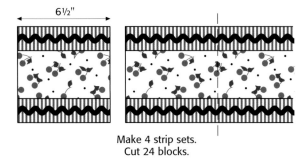

Make 4 strip sets.
Cut 24 blocks.

Assembling the Quilt

1. Lay out the blocks in rows, alternating the blocks as shown.

2. Sew the blocks together in rows, pressing the seam allowances toward the strip-pieced blocks.

3. Sew the rows together, pressing the seam allowances all in one direction.

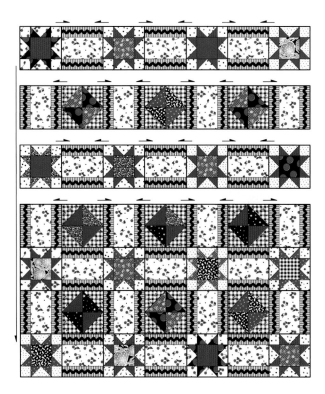

Adding the Borders

1. Measure the length of your quilt, referring to "Adding Borders" on page 8, and trim two of the cherry-print border strips to this length. Sew the border strips to the sides of the quilt and press the seam allowances toward the border strips.

2. Measure the width of your quilt, including the side inner-border strips, and trim the remaining two inner-border strips to this length. Sew the borders to the top and bottom of the quilt and press the seam allowances toward the border.

3. Cut a piece of giant rickrack 2" longer than one side of the cherry-print border. Fold the rickrack in half and mark the center point. Pin the rickrack to the edge of the quilt with the center of the rickrack aligned with the ¼" stitching line as shown. Stitch in place using a thread color that matches the rickrack. Repeat with the remaining three sides.

Center

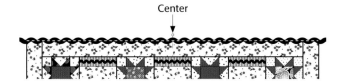

4. Sew the red strips together to make one long strip. Measure the length of your quilt, including the inner border. Cut two red border strips to this length. Sew the border strips to the sides of the quilt and press the seam allowances toward the red border.

Joan's Helpful Hint

As you sew, place the red border strip on the underside and the cherry-print border on top; use the stitching line from the rickrack as a stitching guide.

5. Measure the width of your quilt top, including the red border strips. Cut two red border strips to this length. Sew the border strips to the top and bottom of your quilt and press the seam allowances toward the border.

6. Measure the length of the quilt top and repeat steps 4 and 5 for the black-and-white-check borders.

Finishing the Quilt

Refer to the general directions on pages 9–12 for more details on quilt finishing, if needed.

1. Cut the backing fabric so that it is approximately 4" to 6" larger than the quilt top.

2. Layer the backing, batting, and quilt top, and baste the layers together.

3. Hand or machine quilt as desired.

4. Trim the batting and backing fabric so that the edges are even with the quilt top. Add a sleeve for hanging the quilt, if desired. Bind and label your quilt.

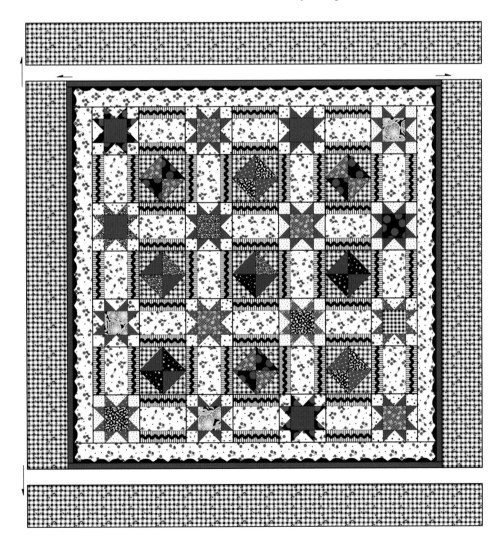

Baby **Churn Dash**

Designed and made by Joan Hanson, 30" x 30".

Use the waste triangles from "Cherry Star Picnic" to stitch up this cheery little Churn Dash quilt. Since half of the leftover triangles were black and the other half were red, I made five of the blocks with black triangles and red squares and four of the blocks with red triangles and black squares. I strip pieced the square units in the block using the same red and the same black fabrics, but more fabrics could be used for a scrappier look or to use up little bits of leftover fabrics.

Finished Quilt Size: 30" x 30" ▲ Finished Block Size: 5" ▲ Number of Blocks: 9

WASTE NOT, WANT NOT

Use waste triangles from "Cherry Star Picnic" on page 57, or cut squares as directed below.

For Churn Dash Blocks
36 pairs of waste triangles
OR
Cut 72 squares, 2½" x 2½".

NOTE: *You will need 20 pairs of black and checked fabric and 16 pairs of red and checked fabric.*

Materials
42"-wide fabric

- ▲ ⅝ yard of black-and-white check for outer border
- ▲ ⅜ yard of light background for blocks and sashing strips
- ▲ ¼ yard or scraps of red print 2 for sashing squares and outer-border corner squares
- ▲ ¼ yard of red-and-white stripe for inner border
- ▲ ⅛ yard of red print 1 for blocks
- ▲ ⅛ yard of black for blocks and inner-border corner squares
- ▲ 1 yard of fabric for backing
- ▲ ⅜ yard of fabric for binding
- ▲ 34" x 34" piece of batting
- ▲ 2½ yards of jumbo rickrack

Cutting

Fabric	Used for	Number to cut	Size to cut	Second cut
Light background	Blocks	2 strips	1½" x 42"	Cut 1 strip into: 9 squares, 1½" x 1½", and 1 strip, 1½" x 26"
	Sashing strips	4 strips	1½" x 42"	24 rectangles, 1½" x 5½"
Red print 1	Blocks	1 strip	1½" x 42"	—
Black	Blocks	1 strip	1½" x 26"	—
	Inner-border corner squares	4 squares	2" x 2"	—
Red print 2	Sashing squares	16 squares	1½" x 1½"	—
	Outer-border corner squares	4 squares	4¼" x 4¼"	—
Red-and-white stripe	Inner border	2 strips	2" x 42"	4 strips, 2" x 19½"
Black-and-white check	Outer border	4 strips	4¼" x 22½"	—

Piecing the Blocks

1. Sew a red print 1 strip to the 1½" x 42" background strip. Press the seam allowance toward the red fabric. Cut 20 segments 1½" wide. Repeat with the black strip and the 1½" x 26" background strip. Cut 16 segments.

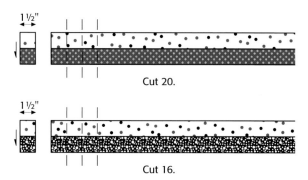

Cut 20.

Cut 16.

2. Referring to "Using Waste Triangles" on page 20, trim pairs of the waste triangles from the Hourglass blocks to 2½" x 2½" on two sides as shown. Using the half-square-triangle technique on page 14, sew the pairs together to create a total of 36 half-square-triangle units. Trim the seam allowances and press toward the black or red fabrics.

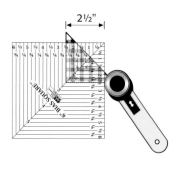

Make 20.   Make 16.

3. Arrange the units from steps 1 and 2 with a 1½" background square as shown. Sew the units together in rows. Press the seam allowances toward the red or black squares. Sew the rows together and press the seam allowances toward the center. Make nine blocks.

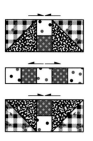 

Make 5 black
and 4 red.

Assembling the Quilt

1. Sew four red print 2 sashing squares and three background sashing strips together as shown. Press the seam allowances toward the red squares. Make four sashing units.

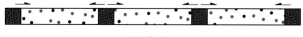

Make 4.

2. Sew four sashing strips and three blocks together as shown. Press the seam allowances toward the sashing strips. Repeat to make three rows.

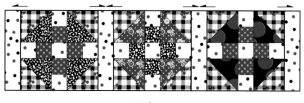

Make 3.

3. Sew the rows and sashing units together. Press the seam allowances toward the sashing.

Adding the Borders

1. Using a 1" seam-allowance guide on your sewing machine, sew the rickrack to the center of the four red-and-white-stripe strips, using a thread color that matches the rickrack.

Make 4.

2. Sew two of the red-and-white-stripe border strips to the sides of your quilt. Press the seam allowances toward the border.

3. Sew a 2" black square to either end of the remaining two red-and-white-stripe borders. Press the seam allowances toward the border.

4. Sew the border strips to the top and bottom of the quilt. Press the seam allowances toward the border.

5. Repeat steps 2 through 4 using the outer-border strips and corner squares. For pressing directions, see the illustration below.

Finishing the Quilt

Refer to the general directions on pages 9–12 for more details on quilt finishing, if needed.

1. Cut the backing fabric so that it is approximately 4" larger than the quilt top.

2. Layer the backing, batting, and quilt top, and baste the layers together.

3. Hand or machine quilt as desired.

4. Trim the batting and backing fabric so that the edges are even with the quilt top. Add a sleeve for hanging the quilt, if desired. Bind and label your quilt.

Dutch Treat **Dishes**

Designed and pieced by Joan Hanson, 66" x 78". Quilted by Dawn Kelly.

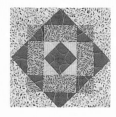

Blue, yellow, and white is such a timeless color combination that it never seems to go out of style. When I was growing up, our kitchen was blue and white and so were our dishes, of course. I am an avid collector of old dishes, and the dish fabric in this quilt reminds me of the set we used for so many years. Notice that very little yellow is used in this quilt; it is mostly an accent color to the blue and white. A little yellow goes a long way! This is a perfect project in which to use fat quarter bundles to give a scrappy look to the blocks.

Finished Quilt Size: 66" x 78" ▲ Finished Block Size: 9" ▲ Number of Blocks: 20

Materials

42"-wide fabric

- ▲ 2¼ yards of blue-and-yellow print for outer border and sashing-star centers
- ▲ 1½ yards of light background for sashing strips
- ▲ 1¼ yards *total* (or 4 fat quarters) of light blue scraps for block background
- ▲ 1 yard *total* (or 4 fat quarters) of assorted dark blue scraps for blocks
- ▲ ⅞ yard of dark blue for sashing-star points

- ▲ ¾ yard *total* (or 4 fat quarters) of assorted medium blue scraps for blocks
- ▲ ½ yard of yellow for inner border
- ▲ ⅜ yard *total* of assorted yellow scraps for blocks
- ▲ 4 yards of fabric for backing
- ▲ ¾ yard of fabric for binding
- ▲ 72" x 84" piece of batting

Cutting

Fabric	Used for	Number to cut	Size to cut	Second cut
Dark blue scraps	Blocks	2 strips	3½" x 42"	20 squares, 3½" x 3½"
		12 strips	2" x 42"	80 squares, 2" x 2"
				80 rectangles, 2" x 3½"
Yellow scraps	Blocks	4 strips	2" x 42"	80 squares, 2" x 2"
Medium blue scraps	Blocks	10 strips	2" x 42"	80 squares, 2" x 2"
				80 rectangles, 2" x 3½"
Light blue scraps	Block backgrounds	18 strips	2" x 42"	80 rectangles, 2" x 3½"
				80 rectangles, 2" x 5"
Dark blue	Sashing-star points	12 strips	2" x 42"	240 squares, 2" x 2"
Light background	Sashing strips	13 strips	3½" x 42"	49 rectangles, 3½" x 9½"
Blue-and-yellow print	Outer border	4 lengthwise strips	6¼" x 70"	—
	Sashing-star centers	2 lengthwise strips	3½" x 70"	30 squares, 3½" x 3½"
Yellow	Inner border	7 strips	2" x 42"	18 rectangles, 2" x 9½"
				22 rectangles, 2" x 3½"
				4 squares, 2" x 2"

Piecing the Blocks

1. Using the square-within-a-square technique on page 17, sew a square cut from the yellow scraps to opposite corners of a 3½" dark blue scrap square. Press the seam allowances toward the yellow fabric. Repeat with the remaining corners. Make 20.

Make 20.

2. Sew a medium blue rectangle to opposite sides of the square units in step 1. Press the seam allowances toward the medium blue rectangles.

3. Sew a dark blue scrap square to both ends of a medium blue rectangle. Press the seam allowances toward the medium blue rectangles. Make 40 units.

Make 40.

4. Sew a unit from step 3 to either side of the center unit from step 2. Press the seam allowances toward the center.

Joan's Helpful Hint

Seams may lie flatter and points look crisper if bulky seams are pressed open.

5. Using the snowball technique on page 15, sew a matching medium blue square to all four corners of the block. Press the seam allowances toward the medium blue fabric.

6. Using the rectangles-with-diagonal-seams technique on page 18, sew a 2" x 3½" light-blue background rectangle to either side of a 2" x 3½" dark blue scrap rectangle. Press the seam allowances toward the dark blue fabric. Make 40 units. Repeat with the 2" x 5" light blue rectangles. Make 40 units.

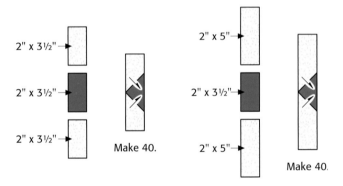

2" x 3½"

2" x 3½"

2" x 3½"

Make 40.

2" x 5"

2" x 3½"

2" x 5"

Make 40.

7. Sew the units to the block centers as shown. Press the seam allowances toward the center or press open if you prefer.

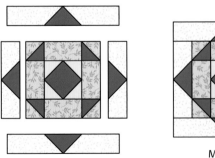
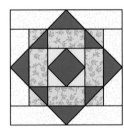

Make 20.

Piecing the Sashing

1. Using the flying-geese technique on page 16, sew a 2" square of dark blue yardage to opposite corners of a light background sashing strip. Press the seam allowances toward the dark fabric. Repeat with the remaining corners. Make 49.

Make 49.

2. Sew four sashing units and five blue-and-yellow-print squares together in a sashing row as shown. Press the seam allowances toward the squares. Make six rows.

Make 6.

Assembling the Quilt

1. Lay out the blocks in rows, adding sashing units. Sew five sashing units and four blocks together as shown. Press the seam allowances toward the blocks. Make five.

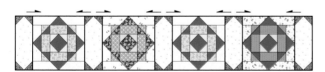

Make 5.

2. Sew the block rows together with the sashing rows as shown. Press the seam allowances in the same direction.

Adding the Borders

1. Using the flying-geese technique, sew two 2" dark blue squares from yardage to a 2" x 3½" yellow rectangle. Press the seam allowances toward the dark blue. Make 22 units.

Make 22.

2. Sew six units from step 1 and five 2" x 9½" yellow rectangles together to make a long border strip. Press the seam allowances toward the blue fabric. Make two long border units for the sides of the quilt.

3. Sew five units from step 1 and four 2" x 9½" yellow rectangles together to make a short border strip. Add a yellow square to each end. Make two short border units for the top and bottom of the quilt.

4. Sew a long border strip to either side of the quilt as shown. Press the seam allowances toward the border strips. Sew the short border strips to the top and bottom of the quilt. Press the seam allowances toward the yellow border.

5. Measure the length of your quilt, referring to "Adding Borders" on page 8, and trim two of the blue-and-yellow border strips to this length. Sew the border strips to the sides of the quilt and press the seam allowances toward the border strips.

6. Measure the width of your quilt top, including the side borders, and trim the remaining two blue-and-yellow border strips to this length. Sew the border strips to the top and bottom of your quilt and press the seam allowances toward the border strips.

Finishing the Quilt

Refer to the general directions on pages 9–12 for more details on quilt finishing, if needed.

1. Piece the backing fabric horizontally and cut it approximately 6" larger than the quilt top.

2. Layer the backing, batting, and quilt top, and baste the layers together.

3. Hand or machine quilt as desired.

4. Trim the batting and backing fabric so that the edges are even with the quilt top. Add a sleeve for hanging the quilt, if desired. Bind and label your quilt.

Country **Village**

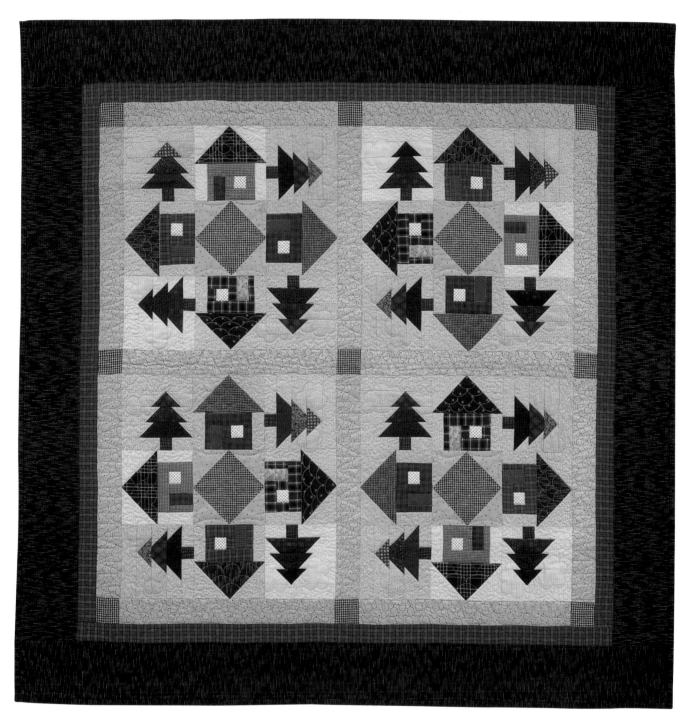

Designed and pieced by Joan Hanson, 55" x 55". Quilted by Dawn Kelly.

A quilt made by Mary Hickey inspired the setting of this quilt. Mary is my dear friend and longtime coauthor. We both love House blocks, and in this quilt I have simplified the House block down to the basics: simple roofline, one door, and one window. The quilt consists of four "villages" made up of four Tree blocks, four House blocks, and a Square within a Square block. Fabrics are muted colors of plaids, stripes, and dots. For the House blocks, I chose four black roof fabrics and four plaid house and door fabrics. For the Tree blocks, I choose four combinations of three green fabrics. For ease of cutting and uniformity, the window fabric and tree trunks are all the same. Each village has one of each house and tree combination, but in different positions in each village. This gives a scrappier look with fewer fabrics for ease of cutting. Fat quarters and scraps work well in this quilt.

Finished Quilt Size: 55" x 55" ▲ Finished Block Size: 6"
Number of Blocks: 16 House blocks, 16 Tree blocks, and 4 Square within a Square blocks

Materials

42"-wide fabric

- ▲ 1½ yards or scraps of tan fabric for block background
- ▲ 1⅛ yards of black for outer border
- ▲ ¼ yard *each* of 4 black prints for roofs
- ▲ ¼ yard *each* of 4 plaids for houses
- ▲ ⅝ yard of tan print for sashing
- ▲ ½ yard of red plaid for inner border
- ▲ ⅛ yard *each* or scraps of 4 plaids for doors
- ▲ ⅛ yard *each* or scraps of 4 greens for trees
- ▲ Fat quarter or ⅜ yard of tan-and-black check for Square within a Square blocks and sashing squares
- ▲ ⅛ yard or scraps of yellow for windows
- ▲ ⅛ yard or scraps of brown for tree trunks
- ▲ 3½ yards of fabric for backing
- ▲ ½ yard of fabric for binding
- ▲ 61" x 61" piece of batting

Piecing the House Blocks

1. Using the flying-geese technique on page 16, sew a 3½" tan square to each side of a 3½" x 6½" black roof rectangle as shown. Press the seam allowances toward the roof fabric. Repeat with all 32 background squares and 16 roof rectangles.

Make 16.

2. Sew a 1" x 11" house rectangle to a door rectangle. Press the seam allowances toward the house fabric. Cut into four segments 2½" wide. Repeat for the remaining three house and door rectangles to make a total of 16 house-door units.

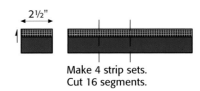

2½"

Make 4 strip sets.
Cut 16 segments.

Cutting

Fabric	Used for	Number to cut	Size to cut	Second cut
Tan	House block backgrounds	3 strips	3½" x 42"	32 squares, 3½" x 3½"
		3 strips	1½" x 42"	32 rectangles, 1½" x 3½"
	Tree block backgrounds	10 strips	1¾" x 42"	16 rectangles, 1¾" x 6½"
				16 rectangles, 1¾" x 4¼"
				16 rectangles, 1¾" x 3¾"
				16 rectangles, 1¾" x 3¼"
				16 rectangles, 1¾" x 2¾"
				16 rectangles, 1¾" x 2¼"
				16 squares, 1¾" x 1¾"
		1 strip	2¼" x 28"	—
		1 strip	3¾" x 28"	—
	Square within a Square block backgrounds	2 strips	3½" x 42"	16 squares, 3½" x 3½"
4 black prints	Roofs; from each fabric, cut:	1 strip	3½" x 28"	4 rectangles, 3½" x 6½" (16 total)
4 plaids	Houses; from each fabric, cut:	4 rectangles	1½" x 3" (16 total)	—
		1 rectangle	1" x 11" (4 total)	—
		2 rectangles	1¼" x 7" (8 total)	—
		4 rectangles	1½" x 4½" (16 total)	—
4 plaids	Doors; from each fabric, cut:	1 rectangle	1½" x 11" (4 total)	—
Yellow	Windows	1 strip	1½" x 42"	4 rectangles, 1½" x 7"
4 green scraps	Trees; from each fabric, cut:	1 strip	1¾" x 14"	4 rectangles, 1¾" x 3" (16 total)
		1 strip	1¾" x 18"	4 rectangles, 1¾" x 4" (16 total)
		1 strip	1¾" x 22"	4 rectangles, 1¾" x 5" (16 total)
Brown	Tree trunks	1 strip	1½" x 28"	—
Tan-and-black check	Square within a Square blocks	1 strip	6½" x 42"	4 squares, 6½" x 6½"
	Sashing squares	1 strip	2½" x 42"	9 squares, 2½" x 2½"
Tan print	Sashing strips	6 strips	2½" x 42"	12 rectangles, 2½" x 18½"
Red plaid	Inner border	5 strips	2" x 42"	—
Black	Outer border	6 strips	5¼" x 42"	—

3. Sew a 1¼" x 7" house rectangle to each side of a window rectangle as shown. Press the seam allowances toward the house fabric. Cut into four segments 1½" wide. Repeat for the remaining house and window rectangles to make a total of 16 window-house units.

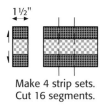

Make 4 strip sets.
Cut 16 segments.

4. Sew the blocks together, pairing up the same roof fabrics with the same house fabric to make the blocks as shown. Press the seam allowances toward the house fabric. Make 16 blocks.

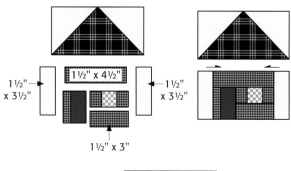

1½"
x 3½"   1½" x 4½"   1½"
                                   x 3½"

1½" x 3"

Make 16.

Piecing the Tree Blocks

1. Using the rectangles-with-diagonal-seams technique on page 18, sew together the background and green rectangles as shown. Press the seam allowances toward the green fabric.

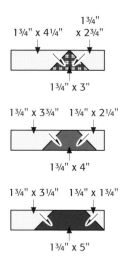

1¾"
1¾" x 4¼"   x 2¾"

1¾" x 3"

1¾" x 3¾"   1¾" x 2¼"

1¾" x 4"

1¾" x 3¼"   1¾" x 1¾"

1¾" x 5"

2. Sew the 2¼" x 28" background strip, the 1½" x 28" brown strip, and the 3¾" x 28" background strip together. Press the seam allowances toward the brown fabric. Cut into 1½" segments. Make 16.

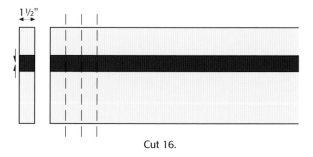

1½"

Cut 16.

3. Sew the blocks together as shown. Press the seam allowances up.

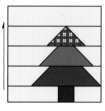

Make 16.

Piecing the Square within a Square Blocks

Using the square-within-a-square technique on page 17, sew a 3½" background square to opposite corners of a 6½" tan-and-black square as shown. Press the seam allowances toward the corners. Repeat with the remaining two corners. Repeat to make four blocks.

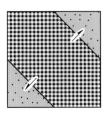 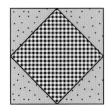

Make 4.

Assembling the Quilt

1. Arrange four different House blocks and four different Tree blocks around a Square within a Square block in the center as shown. Sew together in rows. Press the seam allowances toward the House blocks. Make four village units.

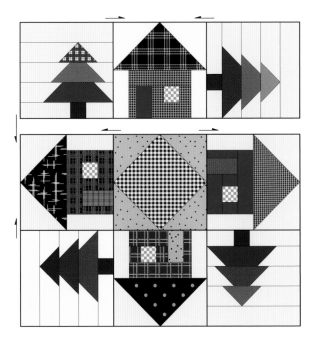

2. Sew three of the 2½" tan-and-black-check squares and two of the tan print sashing strips together as shown. Press toward the sashing strips. Repeat to make three sashing units.

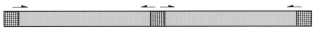

Make 3.

3. Sew three sashing strips and two village units together as shown. Press the seam allowances toward the sashing strips. Repeat to make two rows.

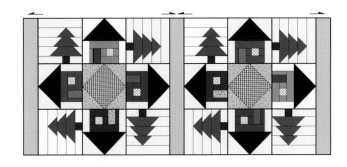

4. Sew the rows together with the sashing units. Press the seam allowances toward the sashing.

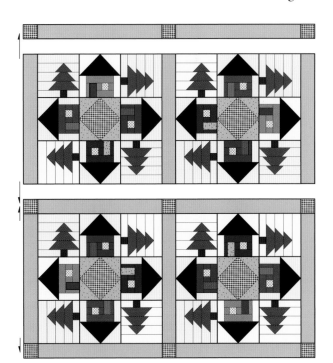

Adding the Borders

1. Measure the length of your quilt, referring to "Adding Borders" on page 8, and trim two of the red inner-border strips to this length. Sew the border strips to the sides of the quilt and press the seam allowances toward the border strips.

2. Cut one of the remaining three red strips in half and sew a half onto each of the other strips. Press the seam allowances open. Measure the width of your quilt top, including the side inner-border strips, and trim border strips to this length. Sew the border strips to the top and bottom of your quilt and press the seam allowances toward the border strips.

3. Repeat steps 1 and 2 for the black outer border.

Finishing the Quilt

Refer to the general directions on pages 9–12 for more details on quilt finishing, if needed.

1. Piece the backing fabric vertically and cut it approximately 4" to 6" larger than the quilt top.

2. Layer the backing, batting, and quilt top, and baste the layers together.

3. Hand or machine quilt as desired.

4. Trim the batting and backing fabric so that the edges are even with the quilt top. Add a sleeve for hanging the quilt, if desired. Bind and label your quilt.

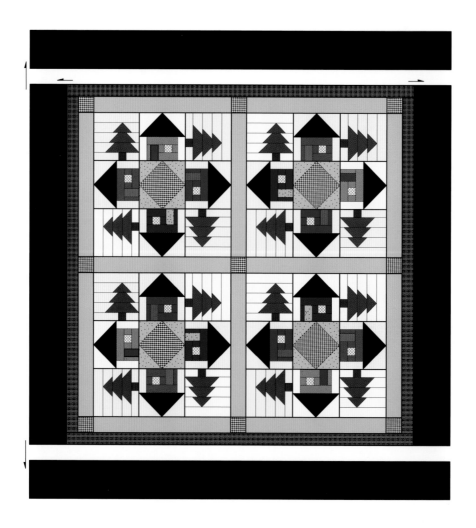

Spring **Flower Garden**

Designed and pieced by Joan Hanson, 40" x 52". Quilted by Dawn Kelly.

Some of my favorite fabrics are the reproduction fabrics from the 1930s. I love the bright clear colors and delightful prints. The flowers in this cheerful quilt are an ideal place to use them. Make the flowers as scrappy as you like or as your fabric collection allows. Each flower uses just a narrow strip of each color. This would be a fun quilt to make as a friendship quilt with each flower made by a different person. If you left the buttons off, it would make a charming baby quilt.

Finished Quilt Size: 40" x 52" ▲ Finished Block Size: 12" ▲ Number of Blocks: 6

Materials

42"-wide fabric

- ▲ 1¼ yards of light print for block backgrounds and inner border
- ▲ ⅞ yard of green print for outer border
- ▲ ¾ yard of assorted prints for flowers
- ▲ ¾ yard of assorted green prints for flower stems, leaves, and bases
- ▲ ½ yard of pink check for middle border
- ▲ Scraps of blue for inner- and outer-border corner squares
- ▲ Scraps of pink for middle-border corner squares
- ▲ 2¾ yards of fabric for backing
- ▲ ½ yard of fabric for binding
- ▲ 46" x 58" piece of batting
- ▲ 18 buttons for flower centers

Joan's Helpful Hint

For ease of cutting, cut two to three flowers out of the same color combination and mix them up in the quilt.

Piecing the Blocks

Making the Flower Units

1. Using the square-within-a-square technique on page 17, for each flower, sew a 1½" square of flower fabric to each corner of a 2½" flower center as shown. Press the seam allowances toward the corners. Make 18.

Make 18.

2. Sew a 1½" x 2½" rectangle of flower fabric to each side of the center unit as shown. Press the seam allowances toward the rectangles.

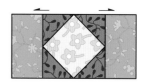

Joan's Helpful Hint

Place the center unit on top and the rectangle on the bottom so that you can see the point in the center of the seam and "aim" for it when stitching.

Cutting

Fabric	Used for	Number to cut	Size to cut	Second cut
Assorted prints	Flowers	18 squares	2½" x 2½" (1 per flower center)	—
		72 squares	1½" x 1½" (4 per flower)	—
		36 rectangles	1½" x 2½" (2 per flower)	—
		36 rectangles	1½" x 4½" (2 per flower)	—
Light print	Block backgrounds	2 strips	4½" x 42"	6 rectangles, 3½" x 4½" 6 rectangles, 2½" x 4½" 6 rectangles, 1½" x 4½"
		6 strips	2" x 42"	12 squares, 2" x 2" 12 rectangles, 2" x 2½" 12 rectangles, 2" x 3" 36 rectangles, 2" x 3½"
		3 strips	1½" x 42"	72 squares, 1½" x 1½"
	Inner border	4 strips	2½" x 42"	—
Assorted green prints	Flower stems	6 rectangles	1½" x 6½"	—
		6 rectangles	1½" x 5½"	—
		6 rectangles	1½" x 4½"	—
	Leaves (Cut 2 from each stem fabric)	36 rectangles	2" x 3½"	—
	Bases	6 rectangles	1½" x 12½"	—
Blue scraps	Inner-border corner squares	4 squares	2½" x 2½"	—
	Outer-border corner squares	4 squares	4¾" x 4¾"	—
Pink check	Middle border	4 strips	2" x 42"	—
Pink scraps	Middle-border corner squares	4 squares	2" x 2"	—
Green print	Outer border	5 strips	4¾" x 42"	—

3. Sew a 1½" x 4½" rectangle of flower fabric to the top and bottom of the flower unit as shown. Press the seam allowances toward the rectangles.

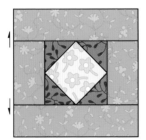

4. Using the snowball technique on page 15, add a 1½" background square to each corner of the flower unit as shown. Press toward the center.

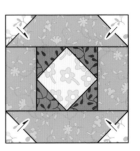

Making the Stem Units

1. Using the rectangles-with-diagonal-seams technique on page 18, sew a 2" x 3" background rectangle to a 2" x 3½" green rectangle as shown. Repeat, adding a 2" background square to the opposite end of the green rectangle as shown. Press the seam allowances toward the green fabric. Make six.

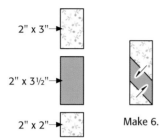

2" x 3"
2" x 3½"
2" x 2"
Make 6.

2. Using a matching green rectangle, repeat step 1 to make the mirror image of the leaf unit. Press the seam allowances toward the green fabric. Make six.

Make 6.

3. Sew the leaf units to either side of a 1½" x 4½" matching green stem rectangle as shown. Press the seam allowances toward the green stem.

1½" x 4½"

Make 6.

4. Repeat steps 1 through 3 to make six each of the following:

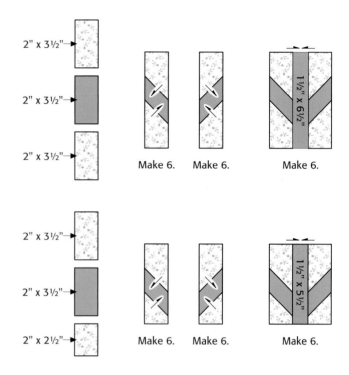

2" x 3½"
2" x 3½"
2" x 3½"
Make 6.   Make 6.   1½" x 6½"   Make 6.

2" x 3½"
2" x 3½"
2" x 2½"
Make 6.   Make 6.   1½" x 5½"   Make 6.

Assembling the Blocks

1. Lay out the flower and stem units in three rows of six flowers in each row as shown. You will have two blocks of three flowers in each row. Each block has a short, tall, and medium stem unit, in that order.

2. When you are happy with the arrangement, add a background rectangle to the top of each flower as shown.

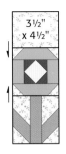 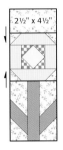 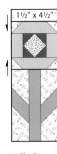

| Short Flower | Tall Flower | Medium Flower |

3. Sew the flower and stem units into blocks as shown. Press the seams toward the flower units. Sew a 1½" x 12½" green rectangle to the bottom of each block as shown. Press the seam allowances toward the green fabric. Make six blocks.

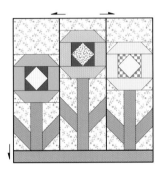

Make 6.

Assembling the Quilt

1. Sew the blocks together in three rows of two blocks each.

2. Sew the three rows together. Press the seam allowances toward the green fabric.

3. Measure the length of your quilt, referring to "Adding Borders" on page 8, and trim two of the background inner-border strips to this length. Measure the width of your quilt and trim the remaining two background strips to this length. Sew the long border strips to the sides of the quilt and press the seam allowances toward the border strips.

4. Sew a blue inner-border corner square to each end of the top and bottom border strips. Press the seam allowances toward the blue fabric. Sew the border strips to the top and bottom of your quilt and press the seam allowances toward the inside.

5. Repeat steps 3 and 4 for the pink middle border and the green-and-blue outer border, piecing the strips as needed.

Quilt Assembly

Finishing the Quilt

Refer to the general directions on pages 9–12 for more details on quilt finishing, if needed.

1. Piece the backing fabric horizontally and cut it 4" to 6" larger than the quilt top.

2. Layer the backing, batting, and quilt top, and baste the layers together.

3. Hand or machine quilt as desired.

4. Sew a button to the center of each flower.

5. Trim the batting and backing fabric so that the edges are even with the quilt top. Add a sleeve for hanging the quilt, if desired. Bind and label your quilt.

Barbie's **Baskets**

Designed and pieced by Joan Hanson, 46½" x 46½". Quilted by Dawn Kelly.

Wouldn't this exuberant quilt make a delightful addition to a little girl's bedroom? Very easy baskets and simple folded flowers are quick and fun to make. Notice how the green fabric adds a calming influence and balance to the hot pink and yellow.

Materials

42"-wide fabric

- ▲ 1½ yards of yellow print for alternate blocks, setting triangles, and outer border
- ▲ 1 yard of light print for Basket block backgrounds
- ▲ ¾ yard of dark pink for Basket blocks, corner squares, and inner border
- ▲ ⅜ yard of green print or scraps for Basket blocks and corner squares
- ▲ ¼ yard of light pink for Basket blocks
- ▲ Assorted scraps of pink and yellow for folded flowers
- ▲ 3 yards of fabric for backing
- ▲ ½ yard of fabric for binding
- ▲ 52" x 52" piece of batting
- ▲ Yellow and pink embroidery floss for flower centers

Cutting

Fabric	Used for	Number to cut	Size to cut	Second cut
Green print	Basket blocks and corner squares	3 strips	2½" x 42"	40 squares, 2½" x 2½" 1 strip, 2½" x 21"
Light print	Basket block backgrounds	6 strips	2½" x 42"	2 strips, 2½" x 14" 36 rectangles, 2½" x 4½" 9 squares, 2½" x 2½"
		1 strip	4½" x 42"	9 squares, 4½" x 4½"
Light pink	Basket blocks	2 strips	2½" x 42"	3 strips 2½" x 14" 8 squares, 2½" x 2½"
Dark pink	Basket blocks and corner squares	3 strips	2½" x 42"	3 strips, 2½" x 14" 10 squares, 2½" x 2½" 1 strip, 2½" x 21"
	Inner border	4 strips	2½" x 36"	—
Yellow print	Alternate blocks and setting triangles	2 squares	13" x 13"	Cut twice diagonally.
		4 squares	8½" x 8½"	—
		2 squares	7" x 7"	Cut once diagonally.
	Outer border	4 strips	4½" x 42"	—
Assorted pink and yellow scraps	Folded flowers	60 to 70 circles	Use patterns on page 85.	—

Piecing the Blocks

1. Using the flying-geese technique on page 16, sew a green square to both sides of a background rectangle. Press toward the green fabric. Make 18 units.

Make 18.

2. Using the 2½" x 14" strips of background, light pink, and dark pink, sew the strips together in the combinations as shown. Press as shown. Cut the strips into 2½" segments as indicated.

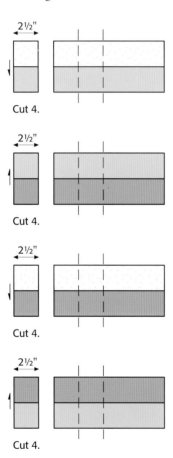

Cut 4.

Cut 4.

Cut 4.

Cut 4.

3. Sew the segments into four-patch units as shown.

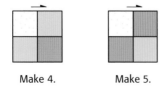

Make 4.   Make 5.

4. Using a 2½" background square, two 2½" x 4½" background rectangles, and two pink squares, piece the block together as shown. Press the seam allowances toward the top of the basket.

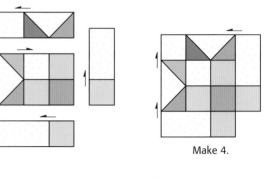

Make 4.

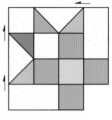

Make 5.

5. Using the big-and-little-triangles technique on page 19, add the 4½" background square last, lining up the sides and pinning the center point as a stitching guide.

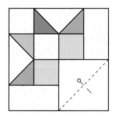

Assembling the Quilt

1. Lay out the Basket blocks with the dark pink blocks in the center and corners and the light pink blocks on the sides. Add the yellow print alternate blocks and setting triangles.

2. Sew the blocks and side triangles together in rows, pressing the seam allowances toward the yellow print.

3. Sew the rows together, adding the 7" corner triangles last. Press the seam allowances toward the yellow fabric.

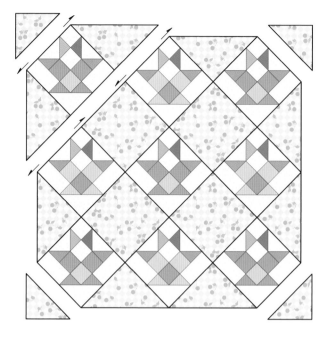

4. Square up the quilt, trimming the edges ¼" from the corners of the blocks.

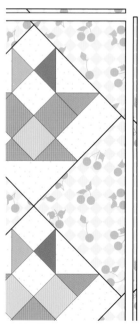

Trim ¼" from block corners.

Embellishing with Flowers

1. From the assorted yellow and pink scraps, cut a variety of 1½", 2", and 2½" circles. You will need about 7 circles per basket, for a total of 63.

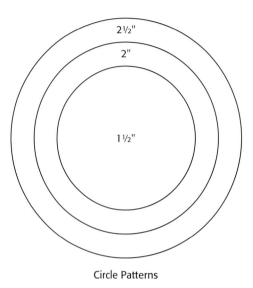

Circle Patterns

2. Press the circles in quarters to locate the center of the circle.

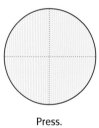

Press.

3. Fold the edge to the center of the circle and press as shown.

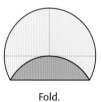

Fold.

4. Fold the next edge to the center and press as shown.

5. Repeat folding and pressing until a hexagon is formed.

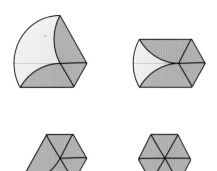

Joan's Helpful Hint

To keep from burning your fingers, use a wooden Popsicle stick to hold the fabric in place while pressing.

6. Arrange seven hexagon flowers on each basket. I placed the folded sides up to give the impression of petals. Pin in place. Using a thread that matches the hexagon, hand stitch to secure the centers, then the outer edges of each flower. Use a contrasting embroidery floss to make three French knots in the center of each flower.

Adding the Borders

1. Measure the length of your quilt, referring to "Adding Borders" on page 8, and trim the four dark pink border strips to this length. Sew two of the border strips to the sides of the quilt and press the seam allowances toward the border strips.

2. Sew the remaining four green squares to the ends of the remaining two dark pink border strips. Sew the border strips to the top and bottom of your quilt and press the seam allowances toward the border strips.

3. Sew the 2½" x 21" dark pink strip and the 2½" x 21" green strip together. Press the seam allowances toward the green fabric. Cut eight segments 2½" wide. Make four four-patch units for the corner squares.

4. Repeat steps 1 and 2 for the yellow outer border. Use the corner squares made in step 3.

Finishing the Quilt

Refer to the general directions on pages 9–12 for more details on quilt finishing, if needed.

1. Piece the backing fabric vertically and cut it approximately 4" to 6" larger than the quilt top.

2. Layer the backing, batting, and quilt top, and baste the layers together.

3. Hand or machine quilt as desired.

4. Trim the batting and backing fabric so that the edges are even with the quilt top. Add a sleeve for hanging the quilt, if desired. Bind and label your quilt.

Kelley's **Quilt**

Designed and pieced by Joan Hanson, 56" x 72". Quilted by Dawn Kelly.

I love this quilt for the optical illusion that it creates. Where does one block end and the next one begin? Is this a Square within a Square block set on point with pieced sashings? Wait a minute. I see purple stars. It's just two alternating blocks with careful placement of the dark and light fabrics that creates this illusion.

Finished Quilt Size: 56" x 72" ▲ Finished Block Size: 8"
Number of Blocks: 18 Sawtooth Star blocks, 17 Square within a Square blocks

Materials

42"-wide fabric

- ▲ 2¼ yards of dark green for Square within a Square blocks and outer border
- ▲ 2 yards of light green for Sawtooth Star and Square within a Square blocks
- ▲ 1 yard of light purple for Square within a Square blocks and inner border
- ▲ ⅞ yard of dark plum for Sawtooth Star points and inner-border corner squares

- ▲ ¾ yard of black for Square within a Square blocks
- ▲ ¾ yard of purple-and-green print for Sawtooth Star centers and outer-border corner squares
- ▲ ½ yard of lavender for Sawtooth Star block corners
- ▲ 3½ yards of fabric for backing
- ▲ ⅝ yard of fabric for binding
- ▲ 62" x 78" piece of batting

Cutting

Fabric	Used for	Number to cut	Size to cut	Second cut
Light green	Star blocks	8 strips	2½" x 42"	72 rectangles, 2½" x 4½"
	Square blocks	8 strips	4½" x 42"	68 squares, 4½" x 4½"
Dark plum	Star points and inner-border corner squares	10 strips	2½" x 42"	148 squares, 2½" x 2½"
Purple-and-green print	Star centers	2 strips	4½" x 42"	18 squares, 4½" x 4½"
	Outer-border corner squares	1 strip	6¼" x 42"	4 squares, 6¼" x 6¼"
Lavender	Star block corners	5 strips	2½" x 42"	72 squares, 2½" x 2½"
Dark green	Square blocks	2 strips	4½" x 42"	17 squares, 4½" x 4½"
	Square blocks	5 strips	2½" x 42"	68 squares, 2½" x 2½"
	Outer border	7 strips	6¼" x 42"	—
Light purple	Square blocks and inner border	10 strips	2½" x 42"	68 squares, 2½" x 2½"
Black	Square blocks	8 strips	2½" x 42"	68 rectangles, 2½" x 4½"

Piecing the Sawtooth Star Blocks

1. Using the flying-geese technique on page 16, sew a 2½" dark plum square to each end of a light green rectangle. Press the seam allowances toward the star points. Make 72 units.

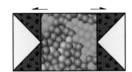

Make 72.

2. Sew a star-point unit to either side of a 4½" purple-and-green-print center square. Press the seam allowances toward the star points. Make 18 units.

Make 18.

3. Sew a lavender square to either side of the remaining star-point units. Press the seam allowances toward the star points. Make 36 units.

Make 36.

4. Sew the units together as shown to make the block. Press the seam allowances toward the center of the block. Make 18.

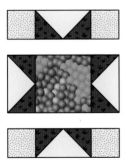 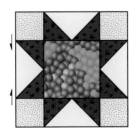

Make 18.

Piecing the Square within a Square Blocks

1. Using the square-within-a-square technique on page 17, sew a light purple square to each corner of a 4½" dark green square as shown. Press the seam allowances toward the light purple. Make 17 center units.

Make 17.

2. Sew a black rectangle to opposite sides of the center unit. Press the seam allowances toward the black fabric. Sew a black rectangle to each of the remaining two sides. Press.

 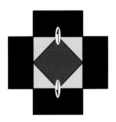

Joan's Helpful Hint

Sew with the center unit facing up so that you can see the point of the square and "aim" for it while stitching.

3. Using the big-and-little-triangles technique on page 19, sew a 4½" light green square to opposite corners of the block. Pin the center point and "aim" for the pin as you sew. Press the seam allowances toward the light green fabric. Repeat with the two remaining corners.

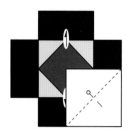

4. Using the snowball technique on page 15, sew a 2½" dark green square to each corner of the block as shown. Press the seam allowances toward the light green. Make 17 blocks.

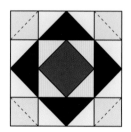 

Make 17.

Assembling the Quilt

1. Lay out the blocks in rows, alternating the blocks.

2. Sew the blocks together in rows, pressing the seam allowances toward the Square within a Square blocks.

3. Sew the rows together, pressing the seam allowances all in one direction.

Adding the Borders

1. Sew the light purple strips together to make a long inner-border strip.

2. Measure the length of your quilt, referring to "Adding Borders" on page 8, and cut two strips to this length. Measure the width of your quilt and trim two strips to this length. Sew the border strips to the sides of the quilt and press the seam allowances toward the border strips.

3. Sew a dark plum corner square to each end of the top and bottom border strips. Press the seam allowances toward the dark plum fabric. Sew the borders to the top and bottom of your quilt and press the seam allowances toward the inside.

4. Repeat steps 1–3 for the dark green outer border with purple-and-green-print corner squares.

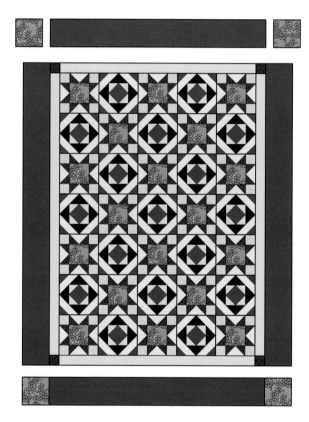

Finishing the Quilt

Refer to the general directions on pages 9–12 for more details on quilt finishing, if needed.

1. Cut the backing fabric so that it is approximately 6" larger than the quilt top.

2. Layer the backing, batting, and quilt top, and baste the layers together.

3. Hand or machine quilt as desired.

4. Trim the batting and backing fabric so that the edges are even with the quilt top. Add a sleeve for hanging the quilt, if desired. Bind and label your quilt.

Little **Shoo Fly**

Designed and made by Joan Hanson, 31¼" x 36¼".

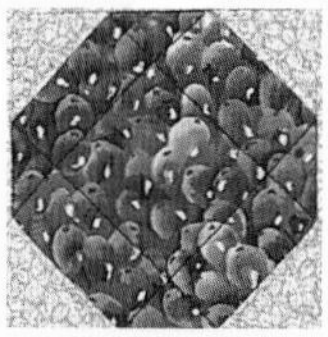

These little Shoo Fly blocks and Snowball blocks are set together in vertical rows with background sashing in between the blocks. The vertical rows are offset to give a honeycomb effect. About a third of the blocks are Snowball blocks and two thirds of the blocks are Shoo Fly blocks. Any combination of these two blocks will work, depending on how much piecing you wish to do.

Finished Quilt Size: 31¼" x 36¼" ▲ Finished Block Size: 3¾"
Number of Blocks: 9 Snowball blocks and 13 Shoo Fly blocks

WASTE NOT, WANT NOT

Use waste triangles from "Kelley's Quilt" on page 87, or cut squares as directed below.

For Snowball Blocks
36 waste triangles
(4 matching triangles for each block)
OR
Cut 36 squares, 1¾" x 1¾".

For Shoo Fly Blocks
52 pairs of light and dark waste triangles
(4 matching pairs for each block)
OR
Cut 104 squares, 1¾" x 1¾" (52 light and 52 dark)

Materials
42"-wide fabric

▲ ⅝ yard of light green for sashing and inner border
▲ ⅝ yard of green print for outer border
▲ ½ yard *total* of medium and dark scraps for blocks
▲ ¼ yard of purple print for middle border
▲ 1¼ yards of fabric for backing
▲ ⅜ yard of fabric for binding
▲ 35" x 40" piece of batting

Cutting

Fabric	Used for	Number to cut	Size to cut	Second cut
Medium and dark scraps	Snowball blocks	9 squares	4¼" x 4¼"	—
	Shoo Fly blocks	65 squares	1¾" x 1¾" (5 per block)	—
Light green	Sashing strips	2 strips	4¼" x 42"	17 rectangles, 1¾" x 4¼" 6 rectangles, 3" x 4¼"
	Inner border	4 strips	1¾" x 42"	—
Purple print	Middle border	4 strips	1½" x 42"	—
Green print	Outer border	4 strips	4¼" x 42"	—

Piecing the Snowball Blocks

1. Referring to "Using Waste Triangles" on page 20, trim the light green waste triangles to 1¾" x 1¾" on two sides as shown. Make 36.

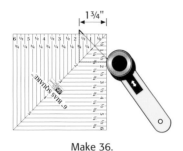

Make 36.

2. Using the snowball technique on page 15, sew a light green waste triangle to each corner of a 4¼" square. Press the seam allowances toward the dark fabric. Make nine Snowball blocks.

Make 9.

Piecing the Shoo Fly blocks

1. Referring to "Using Waste Triangles" on page 20, trim the light and dark pairs of waste triangles to 1¾" x 1¾". Trim four matching pairs for each of the 13 Shoo Fly blocks.

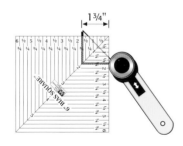

2. Using the half-square-triangle technique on page 14, sew the pairs of waste triangles into half-square triangles. Press the seam allowances toward the dark fabric.

Make 65.

3. Arrange the four matching pairs of half-square triangles and five of the medium and dark 1¾" squares in a pleasing combination. Sew the block together in rows. Press the seam allowances away from the half-square triangles as shown. Make 13 blocks.

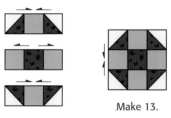

Make 13.

Assembling the Quilt

1. Referring to the quilt diagram, lay out the blocks in vertical rows, adding sashing strips to complete the rows. Sew into rows. Press the seam allowances toward the sashing.

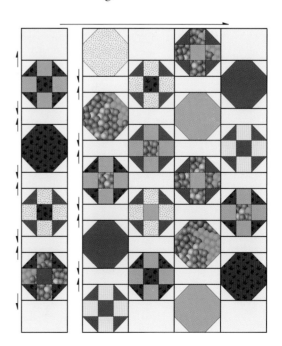

2. Sew the rows together. Press the seam allowances in one direction.

Adding the Borders

1. Measure the length of your quilt, referring to "Adding Borders" on page 8, and trim two of the light green border strips to this length. Sew the border strips to the sides of the quilt and press the seam allowances toward the border.

2. Measure the width of your quilt top, including the side inner-border strips, and trim the remaining two light green border strips to this length. Sew the border strips to the top and bottom of your quilt and press toward the border.

3. Repeat steps 2 and 3 for the middle and outer borders.

Finishing the Quilt

Refer to the general directions on pages 9–12 for more details on quilt finishing, if needed.

1. Cut the backing fabric so that it is approximately 4" larger than the quilt top.

2. Layer the backing, batting, and quilt top, and baste the layers together.

3. Hand or machine quilt as desired.

4. Trim the batting and backing fabric so that the edges are even with the quilt top. Add a sleeve for hanging the quilt, if desired. Bind and label your quilt.

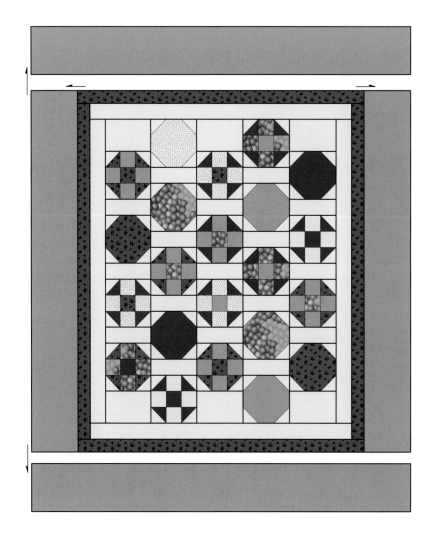

About the **Author**

Joan made her first quilt at the age of 10 for her Barbie doll; this began a lifelong passion for sewing and quiltmaking. She began her teaching career at the high-school level in home economics classes; in more recent years, she has been teaching quilting classes locally as well as around the country. She is the author of several bestselling books, including *Sensational Settings*. She and coauthor Mary Hickey wrote the award-winning book *The Joy of Quilting* and more recently they joined forces to write *The Simple Joys of Quilting*. In addition to quilting, Joan enjoys spending time at her family's cabin on Whidbey Island in Washington State, walking the beach, digging clams, and pulling crab pots.

"My First Quilt," 6" x 7½". Made by Joan Hanson, 1960.

new and bestselling titles from

America's Best-Loved Craft & Hobby Books®

America's Best-Loved Quilt Books®

NEW RELEASES
1000 Great Quilt Blocks
Basically Brilliant Knits
Bright Quilts from Down Under
Christmas Delights
Creative Machine Stitching
Crochet for Tots
Crocheted Aran Sweaters
Cutting Corners
Everyday Embellishments
Folk Art Friends
Garden Party
Hocus Pocus!
Just Can't Cut It!
Quilter's Home: Winter, The
Sweet and Simple Baby Quilts
Time to Quilt
Today's Crochet
Traditional Quilts to Paper Piece

APPLIQUÉ
Appliquilt in the Cabin
Artful Album Quilts
Artful Appliqué
Blossoms in Winter
Color-Blend Appliqué
Sunbonnet Sue All through the Year

BABY QUILTS
Easy Paper-Pieced Baby Quilts
Even More Quilts for Baby
More Quilts for Baby
Play Quilts
Quilted Nursery, The
Quilts for Baby

HOLIDAY QUILTS & CRAFTS
Christmas Cats and Dogs
Creepy Crafty Halloween
Handcrafted Christmas, A
Make Room for Christmas Quilts
Welcome to the North Pole

HOME DECORATING
Decorated Kitchen, The
Decorated Porch, The
Dresden Fan
Gracing the Table
Make Room for Quilts
Quilts for Mantels and More
Sweet Dreams

LEARNING TO QUILT
101 Fabulous Rotary-Cut Quilts
Beyond the Blocks
Casual Quilter, The
Feathers That Fly
Joy of Quilting, The
Simple Joys of Quilting, The
Your First Quilt Book (or it should be!)

PAPER PIECING
40 Bright and Bold Paper-Pieced Blocks
50 Fabulous Paper-Pieced Stars
For the Birds
Quilter's Ark, A
Rich Traditions
Split-Diamond Dazzlers

ROTARY CUTTING
365 Quilt Blocks a Year Perpetual Calendar
Around the Block Again
Around the Block with Judy Hopkins
Fat Quarter Quilts
More Fat Quarter Quilts
Stack the Deck!
Triangle Tricks
Triangle-Free Quilts

SCRAP QUILTS
Nickel Quilts
Scrap Frenzy
Scrappy Duos
Spectacular Scraps
Strips and Strings
Successful Scrap Quilts

TOPICS IN QUILTMAKING
American Stenciled Quilts
Americana Quilts
Batik Beauties
Bed and Breakfast Quilts
Fabulous Quilts from Favorite Patterns
Frayed-Edge Fun
Patriotic Little Quilts
Reversible Quilts

CRAFTS
ABCs of Making Teddy Bears, The
Blissful Bath, The
Handcrafted Frames
Handcrafted Garden Accents
Handprint Quilts
Painted Chairs
Painted Whimsies

KNITTING & CROCHET
365 Knitting Stitches a Year Perpetual Calendar
Clever Knits
Crochet for Babies and Toddlers
Crocheted Sweaters
Knitted Sweaters for Every Season
Knitted Throws and More
Knitter's Book of Finishing Techniques, The
Knitter's Template, A
More Paintbox Knits
Paintbox Knits
Too Cute! Cotton Knits for Toddlers
Treasury of Rowan Knits, A
Ultimate Knitter's Guide, The

Our books are available at bookstores and your favorite craft, fabric, and yarn retailers. If you don't see the title you're looking for, visit us at **www.martingale-pub.com** or contact us at:

1-800-426-3126

International: 1-425-483-3313

Fax: 1-425-486-7596

Email: info@martingale-pub.com

For more information and a full list of our titles, visit our Web site.